Study Guide for CET Examinations

by

J. A. Wilson, CET

&

Dick Glass, CET

HOWARD W. SAMS & CO., INC.
THE BOBBS-MERRILL CO., INC.
INDIANAPOLIS • KANSAS CITY • NEW YORK

FIRST EDITION

THIRD PRINTING – 1972

International Standard Book Number: 0-672-20834-2
Library of Congress Catalog Card Number: 74-143038

Preface

To become a CET, an electronics technician must pass a 12-section exam with a grade of 75% or better. In addition to this, he must have at least four years experience in the electronics servicing industry. This book is not intended to be a complete course in servicing home entertainment electronics. It is aimed primarily at the technician who wants to review the material in a logical sequence before taking the CET exam or a state or local licensing exam. In addition, the material will serve as a review to anyone applying for a position in the electronics field.

Many technicians will not have formal training in some of the areas covered by the exam. For this reason, the chapters are keyed to two television courses published by the Howard W. Sams & Co. Inc. They are the *Photofact® Television Course* and *Color TV Training Manual.* Material for questions on servicing home entertainment equipment other than television is covered in this text. At the beginning of each chapter there are reading assignments in these books so that a technician can obtain a good foundation in the subject matter. The theory portions of this book consist of an in-depth approach to the subject. Special emphasis is placed on more recent developments and uses of such devices as FET's and voltage-variable capacitors.

It is hoped that this book will enable the busy technician to obtain a CET certification and pass a licensing examination where required.

J. A. Wilson
Dick Glass

Contents

1

Scope of the CET Examinations

INTRODUCTION

The purpose of the National Electronic Association's certification program is to provide international recognition for qualified technicians. The program was begun in 1966, and by 1970, 2000 CET (*C*ertified *E*lectronic *T*echnician) certificates, like the one shown in Fig. 1-1, were earned. However, in relation to the total number of technicians elegible to become CET's, only a small percentage of certifications has been issued.

Local and state television servicing organizations have displayed a high interest in the CET program. There is now an *International Society of Certified Electronic Technicians* (*ISCET*). This organization was formed May 16, 1970, at a meeting of the NEA (*N*ational *E*lectronic *A*ssociation) at Lake Charles, Louisiana. Any technician with a CET certificate is elegible to become a member. Fig. 1-2 shows an ISCET membership card.

To become a CET, you must pass a 12-section exam with a grade of 75 percent or better. In addition to this, you must have at least four-years' experience in the electronics servicing industry.

The purpose of this book is to provide you with a broad review of the principles relating to home-entertainment electronics that you may encounter on the CET exam. It is *not* intended to be a complete course in servicing home-entertainment equipment, but a *review* of the material in a logical sequence. This book is also an excellent study guide when preparing for state and local licensing examinations.

There are many technicians who have not had formal training in some of the areas covered by the examinations. For this reason,

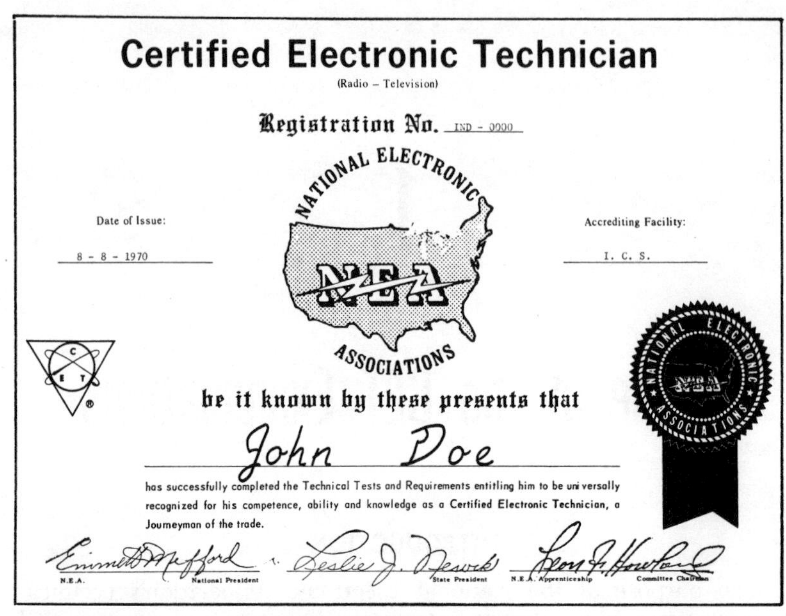

Certified Electronic Technician
(Radio – Television)
Registration No. IND - 0000
NATIONAL ELECTRONIC ASSOCIATIONS
NEA
Date of Issue: 8 - 8 - 1970
Accrediting Facility: I. C. S.
be it known by these presents that
John Doe
has successfully completed the Technical Tests and Requirements entitling him to be universally recognized for his competence, ability and knowledge as a Certified Electronic Technician, a Journeyman of the trade.
N.E.A. National President — State President — N.E.A. Apprenticeship Committee Chairman

Fig. 1-1. A CET certificate.

the chapters in this book are keyed to two television courses published by Howard W. Sams & Co., Inc. They are: *Photofact® Television Course* by The Howard W. Sams Editorial Staff and *Color-TV Training Manual* by The Howard W. Sams Editorial Staff.

At the beginning of each chapter, there are related reading assignments in these books so that a technician can obtain a good foundation in the theory of the material covered in the chapter. The theory discussion sections of this book include additional material related

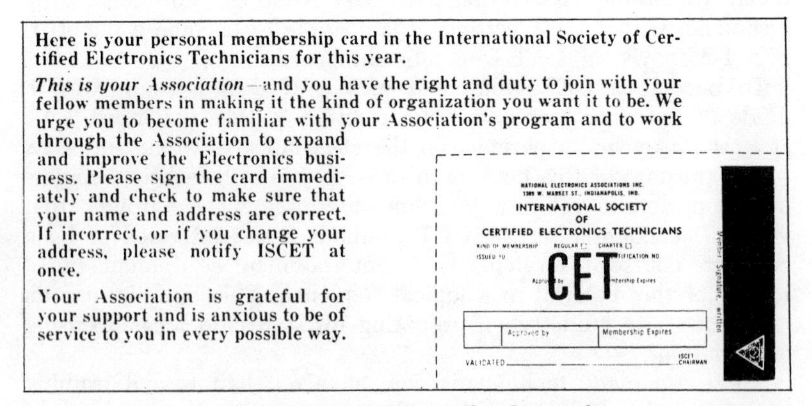

Here is your personal membership card in the International Society of Certified Electronics Technicians for this year.

This is your Association—and you have the right and duty to join with your fellow members in making it the kind of organization you want it to be. We urge you to become familiar with your Association's program and to work through the Association to expand and improve the Electronics business. Please sign the card immediately and check to make sure that your name and address are correct. If incorrect, or if you change your address, please notify ISCET at once.

Your Association is grateful for your support and is anxious to be of service to you in every possible way.

INTERNATIONAL SOCIETY OF CERTIFIED ELECTRONICS TECHNICIANS
CET
Approved by
Membership Expires
VALIDATED
ISCET CHAIRMAN

Fig. 1-2. An ISCET membership card.

to the subject covered in each chapter. Special emphasis is placed on more recent developments and uses of such devices as FET's and voltage-variable capacitors. The CET exam covers both general theory and recent developments.

This study guide is for anyone in the electronics servicing industry who wants to obtain a CET certification or a license. It is also a useful study guide and review for students.

HOW TO USE THIS BOOK

The CET test is divided into twelve sections, each section dealing with a technical subject of importance to radio and TV service technicians. Each chapter of the book deals with one of these subjects. (NOTE: Chapter 6 covers two important areas that are included in the CET test: *Tests and Measurements* and *TV Alignment.*) The written material is on a technical level with the assumption that you are already experienced in the field.

The questions in the tests given at the end of each chapter and in Chapter 12 are NOT the same questions that you will be asked on the CET test administered by the National Electronic Association. However, a score of 75 percent or better in the tests given in this book is an excellent indication that you will be able to pass the CET and licensing examinations.

The following format is used for each chapter.

Keyed Study Assignments

This part of the chapter includes reading assignments in the two Howard W. Sams television course books. You should read this material *before* proceeding with the chapter if you have any doubts about the subject. This way you can obtain a good theory foundation before reading the particular chapter.

Important Concepts

This part of the chapter reviews the important basic concepts that are likely to be encountered in the CET and licensing examinations. It also discusses the newer devices and techniques. This material is presented on a technician's level from the viewpoint of servicing.

Programmed Questions and Answers

This section of the chapter presents programmed questions and answers related to the discussions in the chapter. The purpose is to allow you to review quickly the material in each chapter. This section may also contain questions on the general material in the *Keyed Study Assignments*, but it is primarily for reviewing the material presented in this book. Both the correct and incorrect answers are usually discussed for each question.

Practice Test

This is a 50-question, multiple-choice test on material from both the *Keyed Study Assignments* and the *Important Concepts* sections of each chapter. Answers are given in the back of the book. A grade of 75 percent or better on this test is a good indication that you will be able to pass this subject on the CET and licensing examinations.

Chapter 12 is a practice examination that contains multiple-choice, true-or-false, and word fill-in questions. This practice examination is similar in scope and format to the one given for the certification test. The exception is that the test in Chapter 12 is somewhat longer. A grade of 75 percent or better on this test is a good indication that you can successfully pass the CET and the licensing examinations.

When answering multiple-choice or true-false questions in this book, or on an exam, you may encounter questions in which more than one choice appears to be correct. In such cases, you should pick the answer that is true without qualifications. In other words, pick the answer that is most correct.

Here is an example of a question with more than one correct answer:

Raincoats should be:

(a) yellow.
(b) cooler than bathing suits.
(c) waterproof.

The answer is (c). A raincoat *can* be yellow, and it is (remotely) possible that a bathing suit which is warmer than a raincoat *could* be designed, but the fact that raincoats are supposed to be waterproof is the answer that is best without any qualifications.

Part of the CET exam is devoted to very basic theory. Most experienced technicians do not need to spend time reviewing these basic principles, so this book concentrates on the more advanced theory and practice. (An excellent review of basic principles is given in the Howard W. Sams books titled *ABC's of Electricity* and *ABC's of Electronics.*)

When You Are Ready

When you are ready to take the exam, send a post card with your name and address to: National Electronic Associations, Inc., 1309 West Market Street, Indianapolis, Indiana 46222. Ask for the CET examination schedule for your area. They will send the information to you.

2

The Television Signal

KEYED STUDY ASSIGNMENTS

Howard W. Sams Photofact Television Course
Chapter 9—The Composite Television Signal
Chapter 17—Receiver Controls: Application and Adjustment
Howard W. Sams Color-TV Training Manual
Chapter 1—Colorimetry, The Science of Determining and Specifying Colors
Chapter 2—Requirements of the Composite Color Signal
Chapter 3—Make-up of the Color Picture Signal

This assignment covers both the monochrome and color-TV signals. It also covers the theory of color vision.

IMPORTANT CONCEPTS

If you have been working with television for some time, you no doubt know the frequencies and time intervals that are related to the television signal. However, it will be to your advantage to review them once again as part of your preparation for the CET or license examination.

In addition to the television signal, there is material in this chapter on the vertical interval test signal (VITS) and the fm-stereo multiplex signal.

Television Frequencies and Periods

The horizontal sweep frequency is generally given as 15,750 hertz. This is actually the frequency for a monochrome picture, but it is not exactly the correct frequency for color signals. In order to make room for the color information, the horizontal sweep frequency of the color signal was changed to 15,734.264 hertz. It is usually possible to tell which of the horizontal frequency values is required as an answer in the exam by the way the question is worded, but remember to read each question carefully.

The vertical sweep frequency is usually given as 60 hertz. This is the correct value for monochrome signals, but for color television the correct value is 59.94 hertz. Remember that both the horizontal and the vertical sweep frequencies were *reduced* in order to accommodate the color signals.

The field frequency is 60 hertz, and the frame frequency is 30 hertz. It takes two interlaced fields to make a frame.

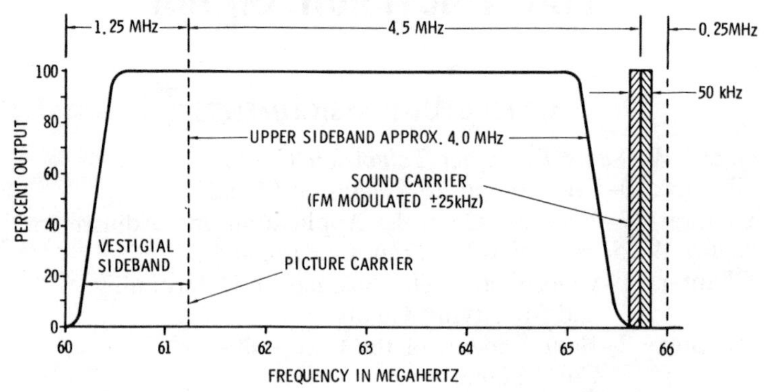

Fig. 2-1. Distribution of frequencies in television Channel 3.

The color-burst frequency is approximately 3.58 megahertz (MHz) above the video carrier. The exact value is 3.579545 MHz (which is slightly lower). The color-burst signal is on the back porch of the horizontal blanking pedestal.

The television channels are 6-MHz wide. The distribution of frequencies in a television channel is shown in Fig. 2-1. It is especially important to note that the sound carrier is 4.5 MHz *above* the video carrier as the signal is *transmitted.* You may be more familiar with the signal frequency distribution in the i-f stage of the television receiver. Fig. 2-2 shows an ideal i-f response curve. In this case the sound carrier is 4.5 MHz *below* the video carrier. The reversal of the frequency distribution takes place in the receiver mixer or converter stage.

The video bandwidth in a color receiver is about 4.2 MHz, but in monochrome receivers the video i-f response may only be 2.5 or 3.0 MHz. This narrow response is at the expense of high video frequencies that produce fine detail in the picture, but a small-

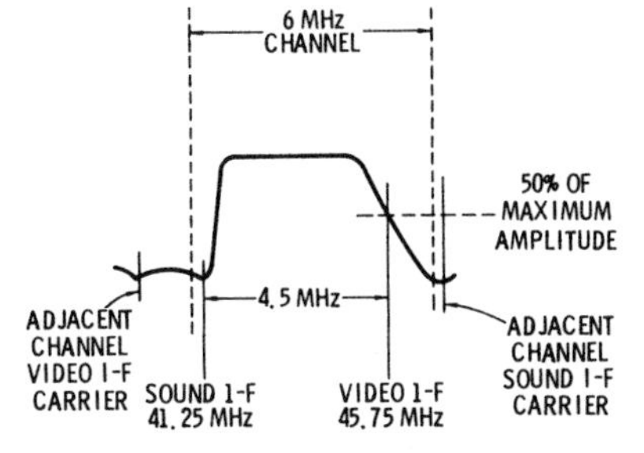

Fig. 2-2. Receiver i-f response curve.

screen portable television receiver will produce an acceptable picture with such a narrow response. A color picture, on the other hand, could not be reproduced in a receiver with a narrow i-f response. This is because the color signals are sent with a subcarrier frequency that is 3.58 MHz above the video carrier. Any reduction in frequency response, then, may result in a loss of color.

The sound i-f frequency in an intercarrier receiver is always 4.5 MHz. (Remember that it is 10.7 MHz in fm broadcast receivers, but 4.5 MHz in television receivers.) This is because the sound i-f signal is obtained by heterodyning (that is, *beating*) the video carrier and sound signals. Since these signals are transmitted with a frequency separation of 4.5 MHz, it follows that they will heterodyne to produce a 4.5-MHz i-f signal. The i-f frequency of a television receiver that is not an intercarrier type may be any frequency. However, such receivers are very rare today.

The time for one line of horizontal scan (H) is approximately 63.5 microseconds. This is an important number because it is used as a standard unit of time for other measurements. For example, the width of the horizontal blanking pedestal is given as 0.18H max. This means that the time for the blanking pedestal cannot be longer than eighteen hundredths of the time allotted for a complete line.

The value 63.5 microseconds is obtained from the basic equation:

$$\text{PERIOD (T)} = \frac{1}{\text{FREQUENCY (f)}}$$

or,

$$\begin{aligned}\text{TIME FOR 1 LINE (H)} &= \frac{1}{\text{LINE FREQUENCY}} \\ &= \frac{1}{15{,}750} \\ &= 63.5\ \mu\text{s (approx.)}.\end{aligned}$$

The value of H includes the time for one horizontal blanking pedestal, one horizontal sync pulse, and one line of video.

The video portion of the composite television signal is amplitude modulated and is transmitted as vestigial sideband. The picture is sent in an aspect ratio (height to width) of 3 to 4. The sound portion of the signal is sent as frequency modulation.

Two color signals are transmitted for color television. Both are transmitted as suppressed-carrier transmission. The *I signal* is sent as a vestigial-sideband, suppressed-carrier wave, and the *Q signal* is sent as a double-sideband, suppressed-carrier wave. To limit the range of frequencies necessary for the transmission of color, the very small areas of a color picture are transmitted as black and white.

Amplitudes

The maximum amplitude of the video signal is from the zero carrier level to the top of the sync pulses. In a *keyed agc* system, it is the horizontal sync-pulse amplitude that determines the amount of agc voltage. Since the amplitudes of the horizontal and vertical sync pulses are equal and not affected by changes in scene brightness, they are a better indication of signal strength than the amplitude of the video signal.

Of the total amplitude, 25 percent is used for the sync pulses. This is shown in Fig. 2-3. Note that the blanking pedestal represents 75 percent of the total amplitude. This is also the maximum amplitude of the video signal. The minimum amplitude of the video signal is 12.5 percent of the maximum amplitude. This minimum value corresponds to the maximum "whiteness" of the picture. The color burst, located on the back porch of the horizontal sync pulse, has an amplitude that is approximately the same as the sync-pulse amplitude (S).

There are three primary colors used for reproducing color-television pictures. They are: *red, green,* and *blue*. In order to produce

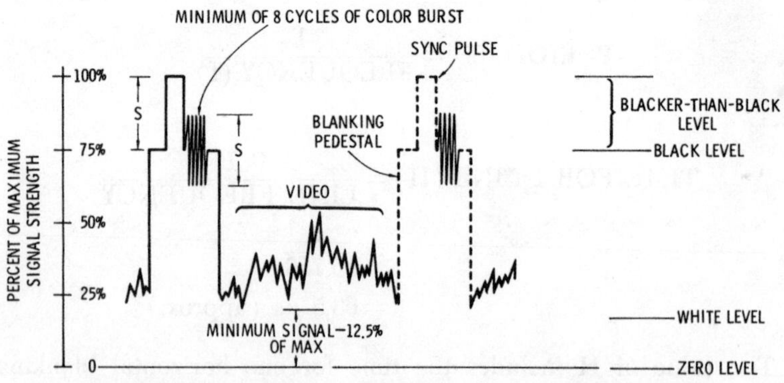

Fig. 2-3. Comparison of modulation levels of a video signal.

white, all three of the primary colors must be combined. If they were combined in equal amplitudes, the human eye would not perceive the resulting color as white. This is because the eye has a different amount of sensitivity to each of the primary colors. To obtain a white color as perceived by the human eye, the primary colors must be mixed in the following proportion: 30 percent red, 59 percent green, and 11 percent blue. The amplitude of the luminance signal is given by:

$$E_Y = 0.30E_R + 0.59E_G + 0.11E_B$$

This is the proportion of voltages on the red, green, and blue guns that produces white on the screen.

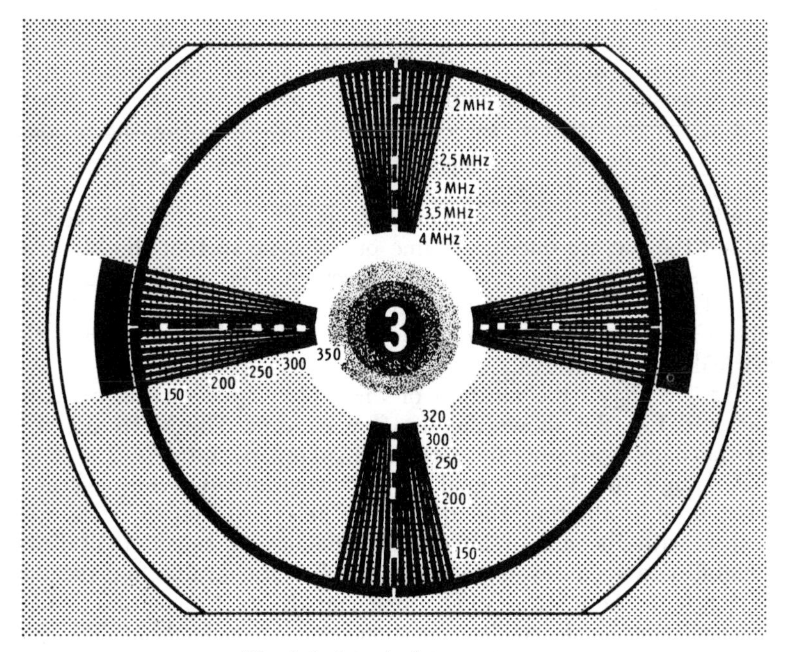

Fig. 2-4. A typical test pattern.

Television Test Signals

Fig. 2-4 shows a typical test pattern. The large dark circle can be used for checking the vertical and horizontal linearity adjustments, the height and width adjustments, and the picture centering.

The horizontal wedges can be used for determining the number of useful scanning lines. The dots of the wedge are numbered according to the number of lines present, and the method is to determine the dot closest to the center at which the lines are still distinguishable. If the lines are distinguishable all the way to the center, then there are at least 350 lines being used.

The dots in the upper vertical wedges are used for determining the frequency response of the receiver. Any ringing that is present will be seen as dark vertical lines immediately following the lines in the vertical wedges.

Poor low-frequency response will be seen as a smear to the right of the black-and-white bars. Insufficient contrast range, or defective agc, is indicated by an inability to obtain five different shades in the concentric circles at the center of the pattern.

Pairing of the interlace lines occurs when the scanning lines are not evenly spaced. This fault is most easily recognized by observing diagonal lines on the test pattern. Pairing causes these lines to be jagged.

Pairing is also indicated by a woven appearance of the lines in the wedges. This effect makes the lines appear like a pattern of coarse weave, and is referred to as the *Moire Effect.*

A test signal that is sometimes transmitted during the vertical interval can be useful for checking the frequency response of color receivers. These transmitted test signals are called *VITS,* which stands for *V*ertical *I*nterval *T*est *S*ignals. They can be observed by rolling the picture one-half frame out of sync so that the blanking bar is seen across the center of the screen. The VITS signal is on lines 17 and 18 of field No. 1 and lines 17 and 18 of field No. 2. These four lines are adjacent to each other and occur during the vertical retrace interval.

Fig. 2-5 shows the location of the VITS signal on the blanking bar, and Fig. 2-6 shows the signal waveform. The VITS signal is on the back porch of the vertical blanking pedestal.

Although the VITS signal is primarily intended for use as a method of evaluating the quality of televised signals, it is also useful, with the aid of an oscilloscope having selection of fields capability and a low-capacitance probe, for evaluating receiver performance. Since the VITS burst signals are transmitted with a constant amplitude at the frequencies shown in Fig. 2-6, they should have a constant amplitude when observed at the first video amplifier. Of course, the 4.2MHz signal *should* be attenuated if the 4.5-MHz traps are properly adjusted. If any of the other burst signals are attenuated, then there is an uneven frequency response in the receiver rf, i-f, or detector stages. The VITS signal can be observed at the first video amplifier stage with the help of a wideband scope that has a triggered sweep.

The VITS signal also includes a staircase pattern with a 3.58-MHz burst on each of the 10 or 11 steps. This part of the signal is more useful for broadcast testing.

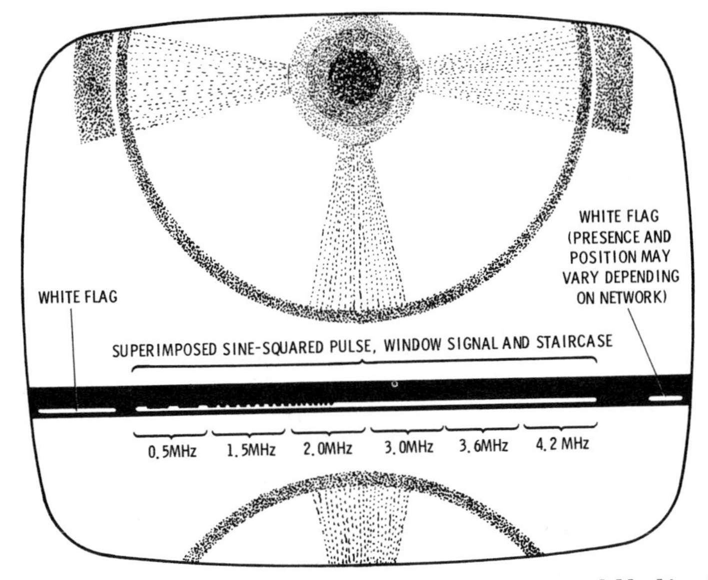

Fig. 2-5. Location of vertical interval test signal on the vertical blanking bar.

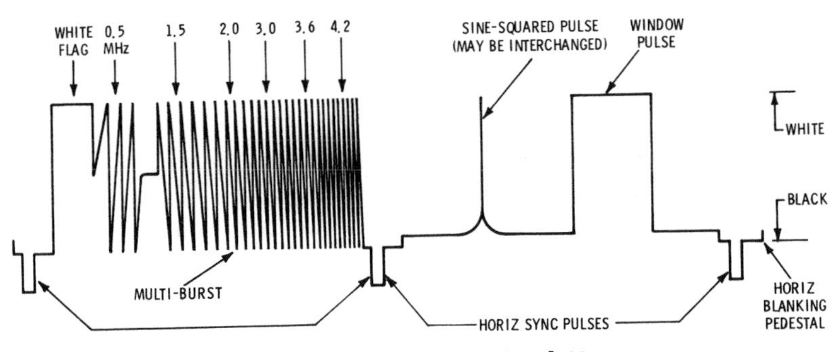

(A) *Field No. 1, lines 17 and 18.*

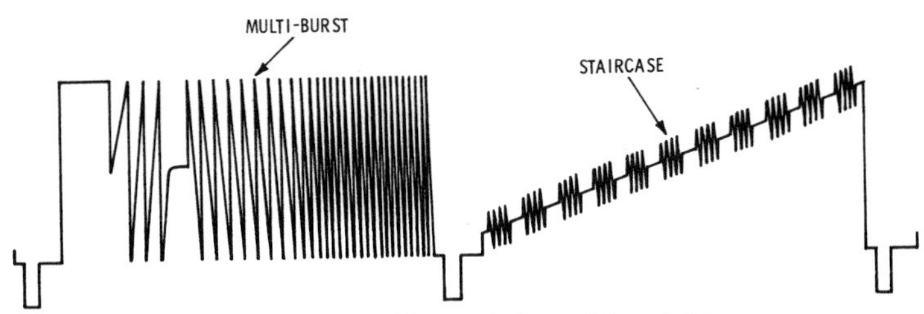

(B) *Field No. 2, lines 17 and 18.*

Fig. 2-6. Vertical interval test signal waveform.

The Stereo Multiplex Signal

The stereo multiplex signal, as illustrated in Fig. 2-7, consists of the following parts:

L + R—This part of the signal contains both right and left sound channels. It is sent as an fm signal, and therefore, an fm receiver can obtain all of the sound when tuned to this signal.

L − R—This is an a-m signal sent with two sidebands and with the carrier suppressed. It is known as the *difference signal.*

Pilot—This is a subcarrier transmitted at 19-kHz above the lower end of the channel. In the receiver, this subcarrier frequency is doubled and used for reinserting the 38-kHz carrier.

SCA (Subsidiary Communications Authorization)—This is an fm signal with a carrier frequency 67 kHz above the lower end of the channel. The SCA signal extends ± 7 kHz around the carrier. This part of the multiplex signal is for background music.

Note that the pilot and SCA signals only modulate the carrier about 10 percent to reduce the possibility of cross talk.

In order to produce two audio channels (L and R) for stereo sound, the L + R and L − R signals are combined as follows:

by addition

$$(L + R) + (L - R) = 2L, \text{ and}$$

by subtraction

$$(L + R) - (L - R) = L + R - L + R = 2R$$

While on the subject of fm signals, remember that the *audio frequency* is related to the number of times per second that the fm signal crosses the center frequency, and the *audio volume* is related to the amount of excursion that the fm signal makes from the center frequency.

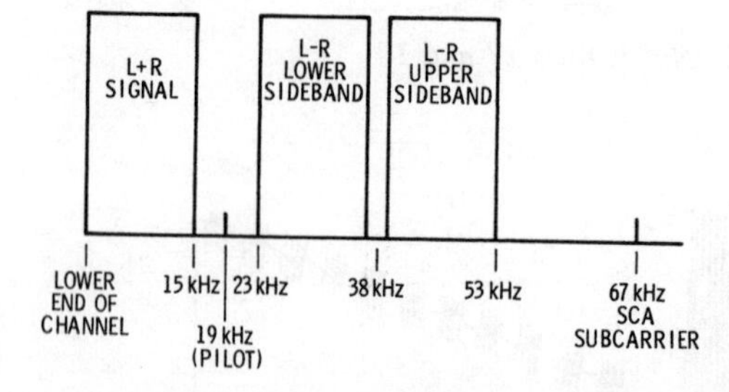

Fig. 2-7. Fm multiplex signal.

PROGRAMMED QUESTIONS AND ANSWERS

Starting with question number 1, select the answer that you feel is correct. If you feel that (A) is correct, proceed to block number

17 as directed. If you feel that (B) is correct, proceed to block number 9 as directed. If you feel that more than one answer is correct, choose the one that you think is the *most* correct.

. . . .

1 When a color signal is being transmitted, the horizontal sweep frequency is:
(A) 15,765.736 hertz. (Go to block number 17.)
(B) 15,734.264 hertz. (Go to block number 9.)

. . . .

2 Your answer is wrong. The video signal (including blanking pedestals and sync pulses) is amplitude modulated and the audio signal is frequency modulated. Go to block number 14.

. . . .

3 The correct answer is pairing of the interlacing.
Here is the next question . . .
A special test signal transmitted during the vertical blanking interval is the
(A) VITS. (Go to block number 18.)
(B) SCA. (Go to block number 7.)

. . . .

4 Your answer is wrong. Narrow-band fm is used for noise-free communications in short-wave systems, but not for fm multiplex. Go to block number 25.

. . . .

5 The correct answer is 63.5 microseconds.
Here is the next question . . .
Which of the following statements is correct?
(A) Only the larger areas of a color television picture are transmitted as color. The very small areas are reproduced in black and white. (Go to block number 22.)
(B) The complete area of a color television picture is transmitted as color. (Go to block number 10.)

. . . .

6 The time for one frame is 33,000 microseconds. The computation is as follows:

$$T = \frac{1}{f}$$

where,
T is the time in seconds,
f is the frequency in hertz.

Thus,

$$T = \frac{1}{30}$$
$$= 0.033 \text{ seconds}$$
$$= 33{,}000 \text{ microseconds.}$$

Here is the next question . . .

Which of the following is correct?

(A) When the amplitude of the video signal is high, the picture-tube screen is dark; and when the amplitude of the video signal is low, the picture-tube screen is light. (Go to block number 19.)

(B) When the amplitude of the video signal is high, the picture-tube screen is light; and, when the amplitude of the video signal is low, the picture-tube screen is dark. (Go to block number 15.)

. . . .

7 Your answer is wrong. The initials SCA stand for *Subsidiary Communications Authorization.* This is part of the fm-multiplex signal. Go to block number 18.

. . . .

8 Your answer is wrong. Go to block number 23.

. . . .

9 The correct answer is 15,734.264 hertz. Remember that the correct value is slightly *below* 15,750 hertz, which is given as the value for monochrome transmission.

Here is the next question . . .

The L − R part of the composite fm-multiplex signal is transmitted as:

(A) narrow-band fm. (Go to block number 4.)

(B) double-sideband, suppressed-carrier a-m. (Go to block number 25.)

. . . .

10 Your answer is wrong. The human eye cannot see small areas of color, so the small areas are transmitted as black and white. Go to block number 22.

. . . .

11 The sync pulses represent 25 percent of the total amplitude, and the blanking pedestal represents 75 percent of the total amplitude.

Here is the next question . . .
In an intercarrier receiver the video carrier is heterodyned (beat) with the fm sound signal to produce the sound i-f. The sound i-f frequency in an intercarrier receiver is:
(A) 10.7 MHz. (Go to block number 8.)
(B) 4.5 MHz. (Go to block number 23.)

. . . .

12 Your answer is wrong. The time for one *horizontal* line is 63.5 microseconds. To find the time for one frame, divide the frame frequency (one-half the field frequency) into one.

$$\text{TIME FOR ONE FRAME (T)} = \frac{1}{\text{FRAME FREQUENCY (f)}}$$

Go to block number 6.

. . . .

13 Your answer is wrong. If you forget the correct value, you can always obtain it from the equation:

$$H = \frac{1}{15{,}750}.$$

Go to block number 5.

. . . .

14 The audio portion of the television signal is frequency modulated. While ±75 kHz is defined as 100 percent modulation for fm broadcast stations, the television (audio) fm signal is 100 percent modulated at ±25 kHz.
Here is the next question . . .
The sync pulses on the composite television signal represent:
(A) 75 percent of the total signal amplitude. (Go to block number 24.)
(B) 25 percent of the total signal amplitude. (Go to block number 11.)

. . . .

15 Your answer is wrong. With *positive picture transmission,* an increase in amplitude corresponds to an increase in brightness. However, the NTSC system employs *negative picture transmission.* Go to block number 19.

. . . .

16 Your answer is wrong. Go to block number 21.

17 Your answer is wrong. The frequency value that you have chosen is slightly *above* the 15,750-hertz value usually given for monochrome signals. Go to block number 9.

18 The correct answer is VITS.

Here is the next question . . .

When an fm receiver is tuned to a multiplex signal, but the fm receiver is not specifically designed for stereo reception, the receiver demodulates the ________ signal. (Go to block number 26.)

19 In the NTSC (National Television Standards Committee) system, negative picture transmission is employed. With negative picture transmission, an increase in amplitude corresponds to an increase in darkness. Remember that the sync tips occur at maximum amplitude and they are in the blacker-than-black region.

Here is the next question . . .

The three primary colors used in color television are red, green, and blue. To produce white, the correct combination of these colors is:

(A) 59 percent red, 11 percent green, 30 percent blue. (Go to block number 16.)

(B) 59 percent green, 30 percent red, 11 percent blue. (Go to block number 21.)

20 Your answer is wrong. Low-frequency phase shift will be represented by streaking or smearing of black lines on the test pattern. Go to block number 3.

21 Color television white is comprised of 59-percent green, 30-percent red, and 11-percent blue. This is apparent from the luminance-signal (also called the Y signal) equation:

$$E_Y = 0.59E_G + 0.30E_R + 0.11E_B$$

Here is the next question . . .

A jagged diagonal line on the test pattern indicates:

(A) low-frequency phase distortion. (Go to block number 20.)

(B) pairing of the interlacing. (Go to block number 3.)

. . . .

22 Only the larger areas of the color television picture are reproduced in color.

Here is the next question . . .

If the field frequency in television transmission is 60 hertz, the period for one frame is:

(A) 33,000 microseconds. (Go to block number 6.)

(B) 63.5 microseconds. (Go to block number 12.)

. . . .

23 The sound i-f frequency is 4.5 MHz. Did the value 10.7 MHz look familiar? It is the i-f frequency used in fm broadcast receivers.

Here is the next question . . .

The various time intervals for the composite television signal are often marked as some multiple of H. The value of H, in microseconds, is:

(A) 18.3 microseconds. (Go to block number 13.)

(B) 63.5 microseconds. (Go to block number 5.)

. . . .

24 Your answer is wrong. Study Fig. 2-3, then go to block number 11.

. . . .

25 The L − R signal is sent as a double-sideband, suppressed-carrier signal.

Here is the next question . . .

The audio portion of the composite television signal is:

(A) frequency modulated. (Go to block number 14.)

(B) amplitude modulated. (Go to block number 2.)

. . . .

26 The L + R signal contains all of the audio. An fm receiver that cannot demodulate a stereo signal will tune to the L + R signal. When fm stereo was first proposed, it was a requirement that all of the receivers in existence could tune to a stereo station without loss of audio information. The L + R signal is the part of the multiplex signal that makes this possible.

You have now completed the programmed questions and answers.

. . . .

PRACTICE TEST

1. Which of the following is not transmitted by amplitude modulating the carrier?

(a) The television video signal.
(b) The sound portion of the television signal.
(c) The vertical blanking pedestal.
(d) The horizontal sync pulse.

2. Fig. 2-8 shows the vertical blanking bar stopped at the center of the picture. Which of the letters marks the VITS signal?

(a) A.
(b) B.
(c) C.
(d) D.

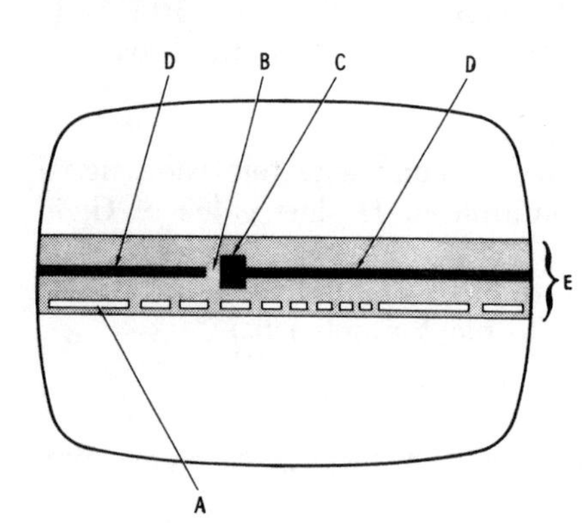

Fig. 2-8. Vertical blanking bar.

3. In Fig. 2-8, which of the letters marks the blanking pedestal?

(a) A.
(b) C.
(c) D.
(d) E.

4. In Fig. 2-8, which of the letters marks the equalizing pulses?

(a) A.
(b) C.
(c) D.
(d) E.

5. In Fig. 2-8, which of the letters marks the serrations in the vertical sync pulse?

(a) A.
(b) B.
(c) D.
(d) E.

6. In Fig. 2-8, which of the letters marks the vertical sync pulses?

(a) C.
(b) D.
(c) E.
(d) A.

7. Fig. 2-9 shows the horizontal blanking bar stopped at the center of the picture. Which of the letters marks the horizontal blanking pedestal?

(a) A.
(b) B.
(c) C.
(d) D.

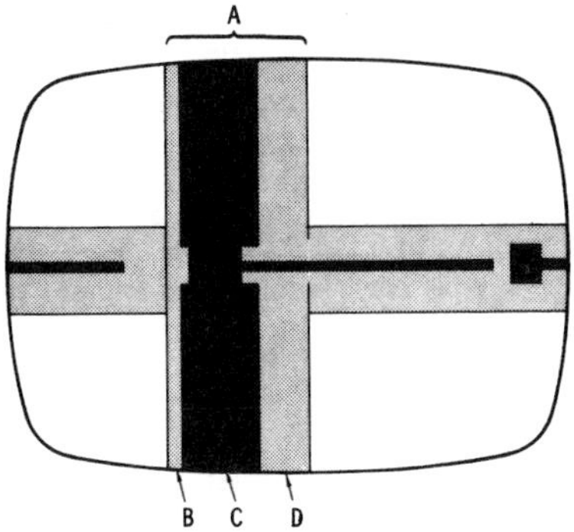

Fig. 2-9. Horizontal blanking bar with vertical blanking bar.

8. In Fig. 2-9, which of the letters marks the front porch?

   (a) A.
   (b) B.
   (c) C.
   (d) D.

9. In Fig. 2-9, which of the letters marks the back porch?

   (a) A.
   (b) B.
   (c) C.
   (d) D.

10. In Fig. 2-9, which of the letters marks the sync pulse?

    (a) A.
    (b) B.
    (c) C.
    (d) D.

11. In the composite television picture, the color burst is a minimum of eight cycles located:

    (a) on the front porch of the vertical blanking pedestal.
    (b) in the 25 kHz guard band at the end of the channel.
    (c) on the back porch of the horizontal blanking pedestal.
    (d) at the top of the horizontal sync pulse.

12. Fig. 2-10 shows a horizontal sync pulse and one line of video. The time for H is:

    (a) 100 microseconds.
    (b) 63.5 microseconds.
    (c) 18.5 microseconds.
    (d) 3.7 microseconds.

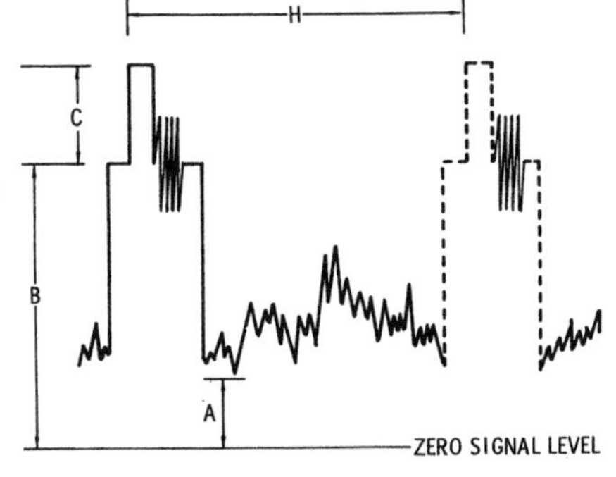

Fig. 2-10. Horizontal sync pulse, blanking pulse, and one line of video

13. **In Fig. 2-10, the minimum amplitude of the video signal, marked as "a" on the drawing, is:**

(a) 0% of maximum.
(b) 7.5% of maximum.
(c) 12.5% of maximum.
(d) 15% of maximum.

14. **In Fig. 2-10, the height of the blanking pedestal is also the height of maximum video signal strength. This is marked *B* on the illustration. What percent of the maximum signal strength is represented by *B*?**

(a) 12.5%
(b) 25%
(c) 50%
(d) 75%

15. **The height of the sync pulse is marked *C* in Fig. 2-10. What percent of the maximum signal strength is represented by *C*?**

(a) 12.5%
(b) 25%
(c) 50%
(d) 75%

16. **Which of the following statements is not true?**

(a) Vertical blanking takes a longer period of time than horizontal blanking.
(b) During vertical retrace, the horizontal lines are still being scanned.
(c) The amplitude of the vertical sync pulse is greater than the amplitude of the horizontal sync pulse.
(d) Although there are 525 lines per frame, only about 480 lines contain video modulation. The rest are lost during vertical retrace.

17. **The type of transmission used with television video signals is:**

(a) amplitude modulation with two sidebands.
(b) amplitude modulation with one complete sideband and one vestigial sideband.
(c) single-sideband, suppressed-carrier transmission.
(d) pulse position modulation.

18. **In the television signal, as it is transmitted:**

(a) the video carrier is at a higher frequency than the center frequency of the sound signal.
(b) the video carrier is at a lower frequency than the center frequency of the sound carrier.

19. **Some of the horizontal lines are lost during vertical retrace. The number of active lines for a frame is:**

(a) 250 to 260.
(b) 470 to 480.
(c) 510 to 515.
(d) 500 to 510.

20. **The maximum deviation of the sound signal for television is:**

(a) 25 kHz (a total carrier swing of 50 kHz).
(b) 75 kHz (a total carrier swing of 150 kHz).
(c) 100 kHz (a total carrier swing of 200 kHz).

21. **The video carrier of a composite signal is:**

(a) 1.25 MHz above the lower end of the channel.
(b) 1.25 MHz below the upper end of the channel.
(c) 0.25 MHz above the lower end of the channel.
(d) 0.25 MHz below the upper end of the channel.

22. The center frequency of the fm sound signal and the carrier frequency of the video signal are:

(a) 45 MHz apart.
(b) 4.5 MHz apart.
(c) 45 kHz apart.
(d) 4.5 kHz apart.

23. The ideal response curve of a television receiver causes the video signal at the carrier to be reduced in amplitude by 50 percent. This is necessary in order to:

(a) reduce cross talk between the video carrier and the sound carrier of the adjacent channel.
(b) reduce cross talk between the video carrier and the sound carrier of the channel being received.
(c) prevent overemphasis of lower video frequencies due to the fact that they are transmitted as a vestigial sideband.
(d) reduce receiver cost.

24. The type of picture polarity used for the NTSC system is:

(a) positive picture polarity.
(b) negative picture polarity.

25. The direct-current component, or *bias,* of the signal corresponds to:

(a) the average illumination of the scene being televised.
(b) the average value of the vertical blanking pedestals.
(c) the average value of the horizontal blanking pedestal.
(d) the average value of the horizontal sync pulses.

26. The width of the television channel is:

(a) 4.5 MHz.
(b) 6 MHz.
(c) 10 MHz.
(d) 1.25 MHz.

27. The aspect ratio of the televised picture—horizontal to vertical—is:

(a) 3 to 4.
(b) 4 to 5.
(c) 4 to 3.
(d) 5 to 4.

28. Which of the following statements is true?

(a) The field frequency is one-half the frame frequency.
(b) The field frequency is 1/500 the horizontal frequency.
(c) The field frequency and the vertical sweep frequency are the same value.
(d) The horizontal sweep frequency is 500 times the vertical sweep frequency.

29. The i-f frequency for a standard a-m broadcast receiver is usually:

(a) 10.7 MHz.
(b) 540 kHz.
(c) 0.455 MHz.
(d) 10.7 kHz.

30. According to the standard luminosity response curve, the human eye is most sensitive to:

(a) violet.
(b) blue.
(c) red.
(d) yellow-green.

31. When the television primary colors are combined in the color picture tube so that:

$$E_Y = 0.59E_G + 0.30E_R + 0.11E_B$$

the result is:
(a) a black screen.
(b) a white screen.
(c) a green color on the screen.

32. The *hue* of a color defines the:

(a) brightness of the color.
(b) wavelength of the color.
(c) amount of white mixed with the color.

33. The reference white used with the NTSC color system is:

(a) illuminant A.
(b) illuminant B.
(c) illuminant C.

34. The subcarrier frequency for color is (approximately):

(a) 6 MHz.
(b) 4.25 MHz.
(c) 3.58 MHz.
(d) 1.25 MHz.

35. The purpose of the color burst is:

(a) to present an amplitude-modulated color signal to the receiver for demodulation.
(b) to supply a rainbow of colors for use in testing the receiver.
(c) to supply bursts of color to the picture tube.
(d) to synchronize the color oscillator in the receiver with the color subcarrier generator at the transmitter.

36. The frame frequency is:

(a) 30 frames per second.
(b) 60 frames per second.
(c) 525 frames per second.
(d) 15,750 frames per second.

37. The line frequency is:

(a) 30 lines per second.
(b) 60 lines per second.
(c) 525 lines per second.
(d) 15,750 lines per second.

38. The field frequency is:

(a) 30 fields per second.
(b) 60 fields per second.
(c) 525 fields per second.
(d) 15,750 fields per second.

39. An advantage of transmitting I and Q signals instead of transmitting R – Y and B – Y signals is:

(a) more realistic blue sky.
(b) more realistic green grass.
(c) more realistic flesh tones.
(d) lower cost.

40. Which of the following signals is transmitted with two complete sidebands?

(a) The I signal.
(b) The Q signal.
(c) The monochrome video signal.
(d) The pilot in the fm multiplex signal.

41. **For video color frequencies above 1.5 MHz,**

(a) only the I signal is transmitted.
(b) only the Q signal is transmitted.
(c) only the Y signal is transmitted.

42. **A change in hue causes:**

(a) a change in the amplitude of the chrominance signal.
(b) a change in the burst frequency.
(c) a change in the phase of the chrominance signal.
(d) no change in the televised signal.

43. **The unmodulated beam on the cathode tube produces the:**

(a) VITS.
(b) SCA.
(c) VARS.
(d) raster.

44. **If the sweep frequency of your oscilloscope is 7.875 kHz, and you are observing the signal at the output of the first video amplifier, you should see:**

(a) two fields, but no blanking pedestals.
(b) two frames including two vertical blanking pedestals.
(c) two lines including two horizontal blanking pedestals.
(d) the Q signal.

45. **For the fm-stereo multiplex signal shown in Fig. 2-11, the SCA subcarrier is marked by the letter:**

(a) A.
(b) B.
(c) C.
(d) D.

46. **For the fm-stereo multiplex signal shown in Fig. 2-11, the pilot is marked by the letter:**

(a) A.
(b) B.
(c) C.
(d) D.

47. **For the fm-stereo multiplex signal shown in Fig. 2-11, the L + R signal is marked by the letter:**

(a) A.
(b) B.
(c) C.
(d) D.

48. **For the fm-stereo multiplex signal shown in Fig. 2-11, the L − R signal is marked by the letter:**

(a) A.
(b) B.
(c) C.
(d) D.

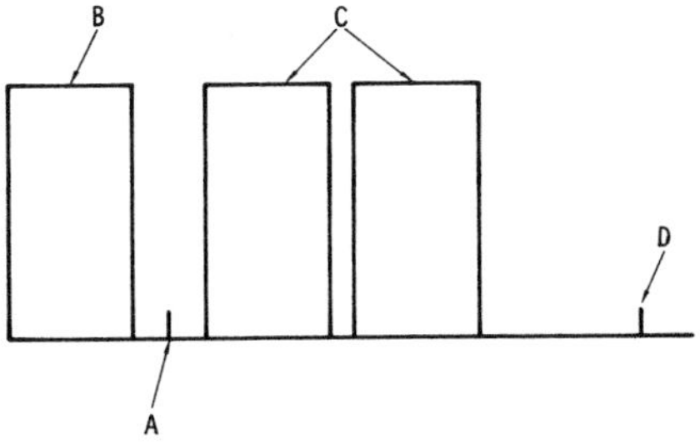

Fig. 2-11. The fm multiplex signal.

49. When the L − R signal is subtracted from the L + R signal, the result is:

(a) no signal. (c) 2R.
(b) 2L. (d) 2L-2R.

50. The i-f frequency of an fm broadcast receiver is:

(a) 455 kHz. (c) 4.5 kHz.
(b) 10.7 MHz. (d) 3.58 kHz.

3

Antennas and Transmission Lines

KEYED STUDY ASSIGNMENT

Howard W. Sams Photofact Television Course
Chapter 11—The Receiving Antenna

This assignment covers antenna theory, types of antennas (including broadband antennas for color reception), and transmission lines.

IMPORTANT CONCEPTS

The Decibel (dB)

The study of antennas and transmission lines always seems to involve work with decibels. If you have been away from textbooks for a number of years, you may feel reluctant to tackle decibels again. However, there are a few shortcuts that will be useful to you when taking the exam. If you know these shortcuts, you can answer the questions involving decibels without working the mathematical equations.

The decibel is a method of comparing two voltages, two currents, or two powers. When you say that an amplifier has a power gain of +3dB, you are saying that the output signal power is greater than the input signal power. On the other hand, a power "gain" of −3dB means that there is a loss of signal power in the amplifier.

If you are mathematically inclined, you can calculate the dB gain or loss by using the following equations:

When comparing two signal powers:

$$dB = 10 \, Log_{10} \frac{P_2}{P_1}$$

When comparing two signal voltages:

$$dB = 20 \, Log_{10} \frac{E_2 \sqrt{R_1}}{E_1 \sqrt{R_2}}$$

When comparing two signal currents:

$$dB = 20 \, Log_{10} \frac{I_2 \sqrt{R_2}}{I_1 \sqrt{R_1}}$$

When the input resistance (R_1) is equal to the output resistance (R_2), they can be eliminated from the equations. When ac impedances are involved, Z_1 is substituted for R_1 and Z_2 is substituted for R_2 in the equations.

Most television service technicians do not carry a table of logarithms around with them, or even a slide rule. These equations are given here to illustrate the following points:

1. When comparing two *powers*, the dB gain or loss involves multiplying the log of the ratios by 10, but for comparing *voltages* or *currents*, the log of the ratio is multiplied by 20.
2. In electronics, the dB gain or loss is *always* obtained by comparing two powers, or two voltages, or two currents.

Tables 3-1 and 3-2 show the ratios that will enable you to work decibel problems in power without using logarithms.

Note that every time that the power is doubled, the power gain is increased by 3 dB and every time the power is halved, the power loss increases by −3 dB.

To solve a decibel problem that involves power, divide the output power by the input power. This gives you the power ratio. Then you will be able to write the dB gain or loss if you know the relationships given in the table. The input power and output power must be in the same units in order to use their ratio to find dB gain or loss.

Table 3-1. Decibel Values for Power Gain

Power Ratio	Power Gain in Decibels
1	0 dB
2	3 dB
4	6 dB
8	9 dB
16	12 dB

Table 3-2. Decibel Value for Power Loss

Power Ratio	Power Gain in Decibels
1	0 dB
½	–3 dB
¼	–6 dB
1/8	–9 dB
1/16	–12 dB

For example, they can both be in microwatts, or both in milliwatts.

Sample Problem—The input power to an amplifier is 200 microwatts, and the output power is 800 microwatts. Fig. 3-1 illustrates the problem. What is the dB gain of the amplifier?

Solution—The ratio of the output power to the input power is $800/200 = 4$. Remember that a ratio of 1 is 0 dB; a ratio of 2 is 3 dB; a ratio of 4 is 6 dB. Now you can write the answer: 6 dB.

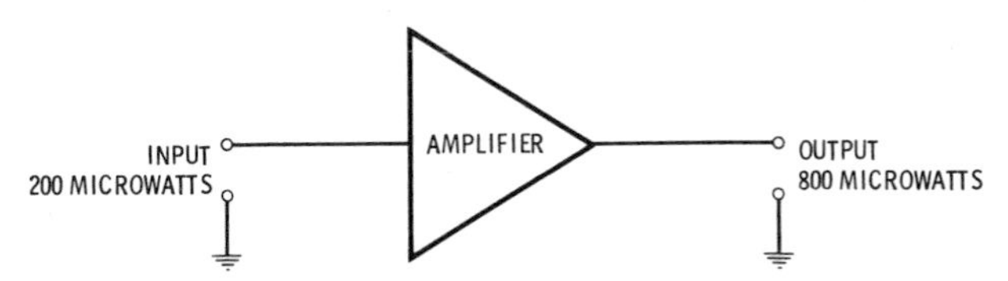

Fig. 3-1. Determine the gain of the amplifier in decibels.

Sample Problem—(See Fig. 3-2.) A transmission line has a loss of 9 dB. If the input power to the line is 800 microwatts, what is the output power?

Solution—A loss of 9 dB corresponds to a ratio of 1/8. This means that the output power will be only 1/8 of the input power.

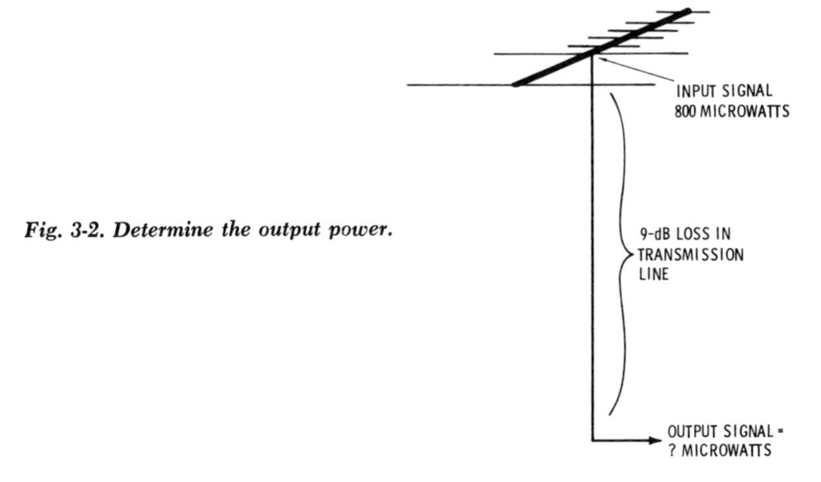

Fig. 3-2. Determine the output power.

Therefore:

$$P = \frac{1}{8} \times 800 = 100 \text{ microwatts.}$$

When calculating dB gain or loss for voltages or currents, you should remember that the log of the ratio is multipled by 20 (instead of by 10 as for the power equation). Therefore, the dB gain or loss is twice as great for a given ratio. This is shown in Tables 3-3 and 3-4.

Table 3-3. Decibel Values for Voltage (or Current) Gain

Voltage (or Current) Ratio	Voltage (or Current) Gain in Decibels
1	0 dB
2	6 dB
4	12 dB
8	18 dB
16	24 dB

Note that *doubling* the voltage (or current) ratio *increases* the gain by 6 dB, and *halving* the voltage (or current) ratio *decreases* the gain by 6 dB. A power ratio of 2 to 1 means a 3-dB increase; but, a voltage (or current) ratio of 2 to 1 means a 6-dB increase.

Sample Problem—A 2000-microvolt signal is delivered by an antenna to a transmission line, but when it reaches the receiver the signal is only 1000 microvolts. Fig. 3-3 illustrates the problem. What is the dB loss of the transmission line system?

Solution—The ratio of voltages is the output of the transmission line (at the receiver) divided by the input (at the antenna):

$$\frac{\text{Output voltage}}{\text{Input voltage}} = \frac{1000}{2000} = \frac{1}{2}.$$

This represents a loss of 6 dB, or a "gain" of −6 dB. In other words, a loss of 6 dB means that half of the signal voltage is lost in the transmission line system. If the problem involves ratios other than the simple ones given here, then you will have to estimate the answer, or else work it the long way with a slide rule. Suppose, for example, that the power ratio is 3. From your table of values you know that

Table 3-4. Decibel Values for Voltage (or Current) Loss

Voltage (or Current) Ratio	Voltage (or Current) Gain in Decibels
1	0 dB
½	−6 dB
¼	−12 dB
1/8	−18 dB
1/16	−24 dB

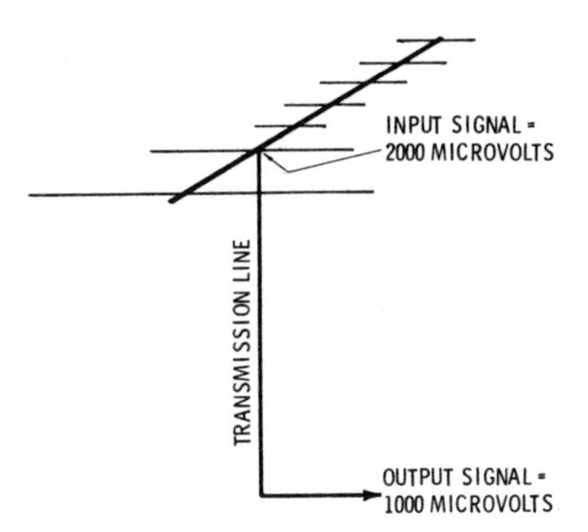

Fig. 3-3. Determine the loss in decibels.

the dB gain is greater than 3 dB and less than 6 dB. (The actual value is 4.77 dB.)

If you understand the basic decibel relationships discussed here, you should have no trouble with decibel problems on examinations.

The Signal

The characteristics of the television signal were reviewed in Chapter 2. Those characteristics that are related to the antenna performance and dimensions are summarized here.

Television antennas are designed for reception in the vhf and uhf bands. By definition, the vhf band extends from 30 to 300 MHz, and the uhf band extends from 300 to 3000 MHz. The fm broadcast signals (88-108 MHz) are in the vhf band between television Channels 6 and 7. The actual frequencies involved are given in Table 3-5.

Table 3-5. Frequency Bands of FM and TV Signals

Channel	Frequency	Band
2-6	54-88 MHz	VHF
FM Broadcast	88-108 MHz	VHF
7-13	174-216 MHz	VHF
14-83	470-890 MHz	UHF

The vhf signals are horizontally polarized. This means that the *electric field* of the wave is horizontal. In order to receive this type of wave, the antenna elements should be horizontal.

Some of the advantages of horizontal polarization are:

1. The horizontal antenna elements are least receptive to vertically polarized noise signals. Since most of the man-made interference is vertically polarized, this means that there will be less noise delivered to the receiver.
2. A horizontally polarized wave will follow the ground with less loss than a vertically polarized wave.

3. Antennas designed for horizontally polarized waves can be made more directive. The narrow horizontal pattern is helpful in the elimination of ghosts.
4. There is better space-wave coverage.

It might be well to review the meaning of the term *space wave.* When a radio wave leaves an antenna, part of the wave moves along the ground to form the *ground wave* and the rest of the wave moves upward and outward to form the *sky wave.*

The ground wave consists of two parts: the *surface wave* and the *space wave.* The surface wave travels along the surface of the earth. The space wave moves immediately above the surface of the earth directly from the transmitting antenna to the receiving antenna or from the transmitting antenna to ground, and from there it is reflected to the receiving antenna. Fig. 3-4 shows the two components of the space wave.

Fig. 3-4. The two components of a space wave.

When the transmitting and receiving antennas are both close to ground, the two components of the space wave cancel because the reflected wave undergoes a phase shift of 180°. When the antennas are several wavelengths above ground, as in the case of television, better coverage is obtained with horizontally polarized waves.

The polarization of uhf signals is a more complicated matter. Such things as weather conditions, the distance from the transmitter, and the amount of vegetation in the region of the receiving antenna can affect the polarization.

In order to be able to deliver the same amount of power to a receiving antenna, a television station operating at a higher frequency must transmit with greater power than a transmitter operating at a lower frequency. The problem of increased losses at higher frequencies is at least partially offset by the fact that the receiving antenna can be designed with a higher gain. This is because the size of the elements is inversely related to the frequency. In the uhf band the elements are quite small, so a large number of directive elements can be used if necessary. Fig. 3-5 shows a typical high-gain uhf antenna. By way of contrast, an elaborate antenna system for reception of Channel 2 would be heavy and bulky.

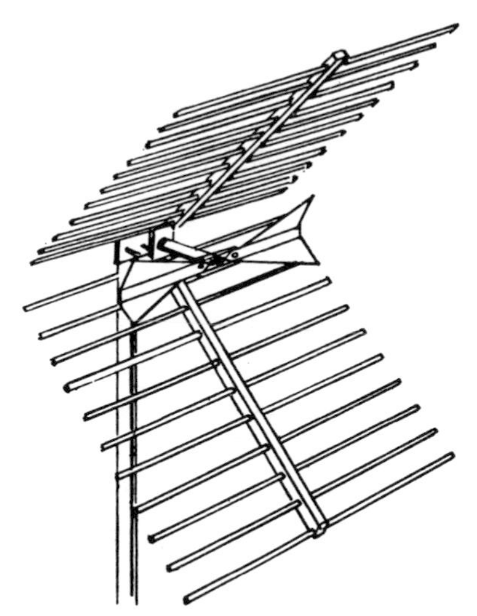

Fig. 3-5. A uhf antenna.

The Antenna

An important characteristic of a television antenna is its *gain*. This term is misleading because it seems to imply that the antenna acts like an amplifier to increase the strength of the received signal. This, of course, is not a true picture of what the antenna does.

An isotropic antenna is one that radiates or receives equally well in all directions. No antenna is truly isotropic, but the theoretical signal strength from such an antenna can be calculated mathematically. The gain of an antenna is a comparison of the amount of signal it receives compared to the amount of signal that an isotropic antenna would receive in the same position.

Instead of using an isotropic antenna as a reference, a simple dipole may be used. Then the *gain* of an antenna is defined as the amount of signal it receives compared to the amount of signal that a dipole antenna receives when mounted in the same position. Since a dipole antenna is bidirectional, it must be turned in the direction that enables it to generate the maximum signal across its terminals when it is used for comparison with other antennas.

Since the gain of an antenna is determined by *comparing* the amount of signal voltage it generates at its terminals with the amount of signal voltage that an isotropic (or dipole) antenna would generate, it is logical to use decibels.

A simple dipole, without directors or reflectors, has a gain of about 2 dB over an isotopic antenna. The higher the dB rating of an antenna, the greater the ability it has to capture a signal. As a general rule, reflectors and directors increase the gain of an antenna.

The *directivity* of an antenna is a measure of its ability to receive a signal from one direction and reject signals from other directions. The horizontal directivity of an antenna is very important for rejecting multipath signals which cause ghosts.

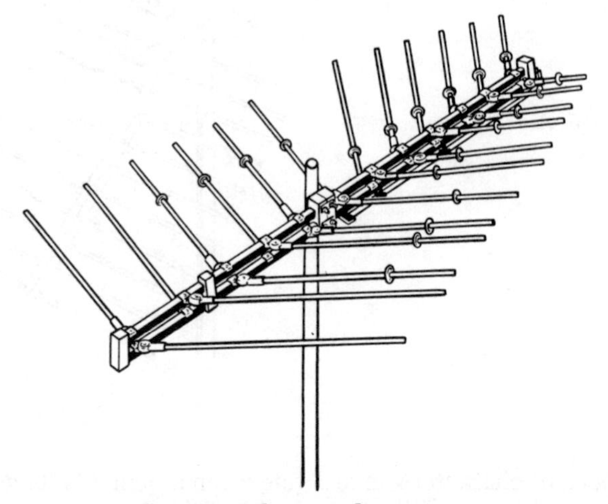

Fig. 3-6. A log-periodic antenna.

If an antenna is to be used for color-television reception, it is very important that it have a wide bandwidth and a flat response. An uneven response can result in poor color reception even though the signal is sufficiently strong for monochrome reception. A log-periodic antenna, such as the one shown in Fig. 3-6, can be designed to receive all channels in the vhf and uhf bands with a bandwidth that is satisfactory for operating a color-television receiver.

When the television receiver is in a location where the transmitters are in different directions, then an antenna rotator may be used to turn the antenna toward the desired station. Instead of using a rotator, two or more antennas can be used. If they are mounted on the same mast, it is important that they be mounted *at least* a half-wavelength apart at the lowest frequency. This is necessary in order to reduce the likelihood of interference between the antennas.

The Transmission Line

The four most important types of transmission lines used in television installations are *flat twin lead, foam-filled twin lead, coaxial cable,* and *shielded twin lead.* Fig. 3-7 shows these transmission lines for comparison.

Flat twin lead has the least amount of loss when it is new. However, as it ages, its losses increase. It is very susceptible to losses when foreign materials are deposited on its surface. For example, painting the wire with a lead-base house paint renders it useless. Moisture and deposits from smoke and smog will also increase its loss to the point where it is not usable.

When twin lead is installed, long horizontal runs should be avoided. There are two reasons for this. First, moisture and dirt are more likely to accumulate on horizontal lines. Second, the horizontal line is more likely to pick up the horizontally polarized signal and produce ghost images on the screen.

If staples are used to hold the line in place, NEVER put the staple *across* the line. Also, the line should never be run near metal objects, such as pipes or metal rain gutters. Nearby metal objects increase the loss of the line.

Considering the disadvantages of twin lead, it is less preferred on installations for color-TV receivers. Foam-filled twin lead has a relatively low loss and is much less affected by moisture and impurity deposits.

Coaxial cable is used because of the fact that it shields the inner conductor from noise signals. Unlike twin lead, it can be installed

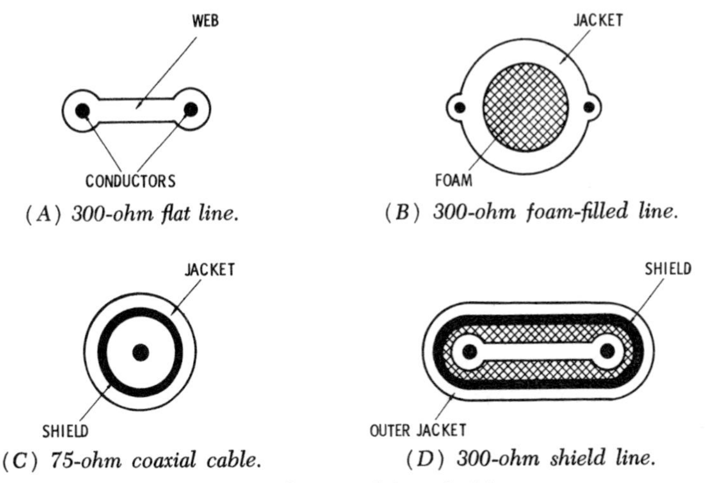

(*A*) *300-ohm flat line.*

(*B*) *300-ohm foam-filled line.*

(*C*) *75-ohm coaxial cable.*

(*D*) *300-ohm shield line.*

Fig. 3-7. Transmission lines used for television antennas.

inside of metal pipes and near metal objects. This makes it useful in MATV (Master Antenna TV) installations. Another advantage is that it does not radiate its signal, and therefore it causes a minimum of interference between installations. Also, impurities on the surface of coaxial cable do not affect it as in the case of twin lead.

The fact that coaxial cable has a low impedance—usually about 75 ohms—is sometimes considered to be a disadvantage, although some receivers have provision for both a low- and a high-impedance input. A special transformer, called a *balun,* can be used to match a 300-ohm line to a coaxial cable. The name comes from the fact that a coaxial cable is unbalanced to ground, and twin lead is balanced to ground. The balun matches the BALanced line to the UNbalanced line.

Shielded 300-ohm lead has the same advantages as coaxial cable, such as low noise pickup and the fact that it will not radiate interference. It is unaffected by moisture and dirt, and can be mounted near metal objects without loss. It has the further advantage that matching transformers are not required for connecting the cable to 300-ohm input or output terminals. The most important disadvantage is that its cost per foot is higher than for other types of transmission line.

The impedance of a transmission line is an important factor to consider when making an installation. When the impedances are not correctly matched, standing waves result. Standing waves always represent a loss because they mean that all of the signal is not being used at the load. In other words, the unused energy is reflected along the line to produce the standing waves.

Auxiliary Systems

There are cases where it is desirable to operate a number of television receivers from one antenna. Systems for doing this range from a simple two-set coupler to an MATV system capable of driving hundreds of receivers.

An MATV system consists of two sections: the *head end* and the *distribution system.* The parts of the head end are shown in Fig. 3-8. The distribution system includes trunk lines (that is, main lines), feeder lines, and outlets to the individual receivers. Traps and filters are used for eliminating undesired signals. Attenuators are used for reducing the signal strength to prevent overdriving the receiver.

The standard of signal strength in MATV systems is 1000 microvolts. The dB value of gain or loss is always obtained by comparing the signal in question with a signal having a strength of 1000 microvolts.

Sample Problem—The preamplifier in an MATV system has a gain of 12 dB. What does this mean?

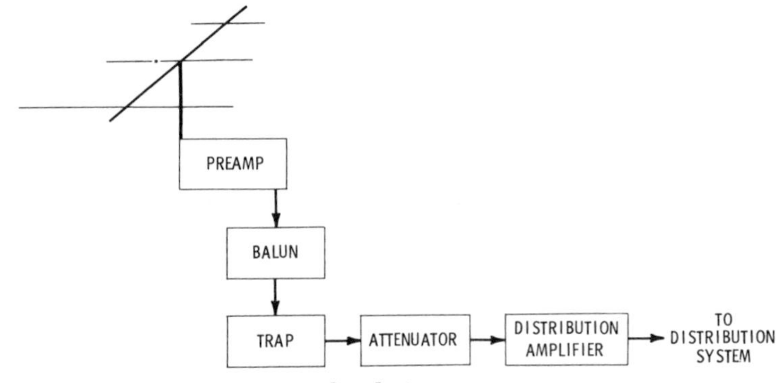

Fig. 3-8. Head end of an antenna system.

Solution—To calculate the gain, you must start with the assumption that the input to the amplifier is 1000 microvolts.

If the output of the amplifier were 1000 microvolts, there would be no gain. The voltage ratio would be:

$$\frac{1000}{1000} = 1.$$

This voltage ratio corresponds to 0 dB.

If the output is 2000 microvolts, then the voltage ratio is:

$$\frac{2000}{1000} = 2.$$

This corresponds to a gain of 6 dB.

If the output is 4000 microvolts, then the voltage ratio is:

$$\frac{4000}{1000} = 4.$$

This corresponds to a gain of 12 dB.

Note that the output of 4000 microvolts will only occur if the input is a standard 1000 microvolts.

Remember this important thing: The calculation of decibel gain or loss in an antenna system is based on a standard voltage of 1000 microvolts or 1 millivolt. This is called dBmV (decibels referred to 1 millivolt).

Antenna Rotators

Two different kinds of rotators are used for television antennas: *manual* and *fully automatic*. With a manual rotator, the antenna turns only while the switch on the unit is in the energized position. This type of rotator has an indicator to show the direction the antenna is pointed.

With an automatic rotator, there are preset positions on the control unit. Pushing one of the preset switches, or turning an indicator knob on some units, causes the antenna to turn to a given position. When the antenna turns to the preset position, it automatically stops turning.

PROGRAMMED QUESTIONS AND ANSWERS

Starting with question number 1, select the answer that you feel is correct. If you feel that (A) is correct, proceed to block number 17 as directed. If you feel that (B) is correct, proceed to block number 9 as directed. If you feel that more than one answer is correct, choose the one that you think is the *most* correct.

. . . .

1 A certain preamplifier doubles the input signal voltage. The voltage gain of this amplifier is:

(A) 3 dB. (Go to block number 17.)

(B) 6 dB. (Go to block number 9.)

. . . .

2 The answer 250 microvolts is right. A loss of 12 dB means that the voltage is halved, and then halved again.

Here is the next question . . .

When two antennas are mounted on the same mast,

(A) they should be mounted as close to each other as possible so that they will help each other. (Go to block number 19.)

(B) they should be mounted as far apart as possible, and never closer than a half wavelength for the lowest channel to be received. (Go to block number 12.)

. . . .

3 The use of horizontally polarized waves is *helpful* in eliminating ghosts because the horizontal polar pattern can be made narrower than a vertical polar pattern.

Here is the next question . . .

Standing waves on a transmission line are:

(A) caused by an impedance mismatch. (Go to block number 6.)

(B) desirable if the signal strength is high. (Go to block number 8.)

. . . .

4 Your answer is wrong. For maximum transfer of power between the television transmitter (which transmits a horizontally polar-

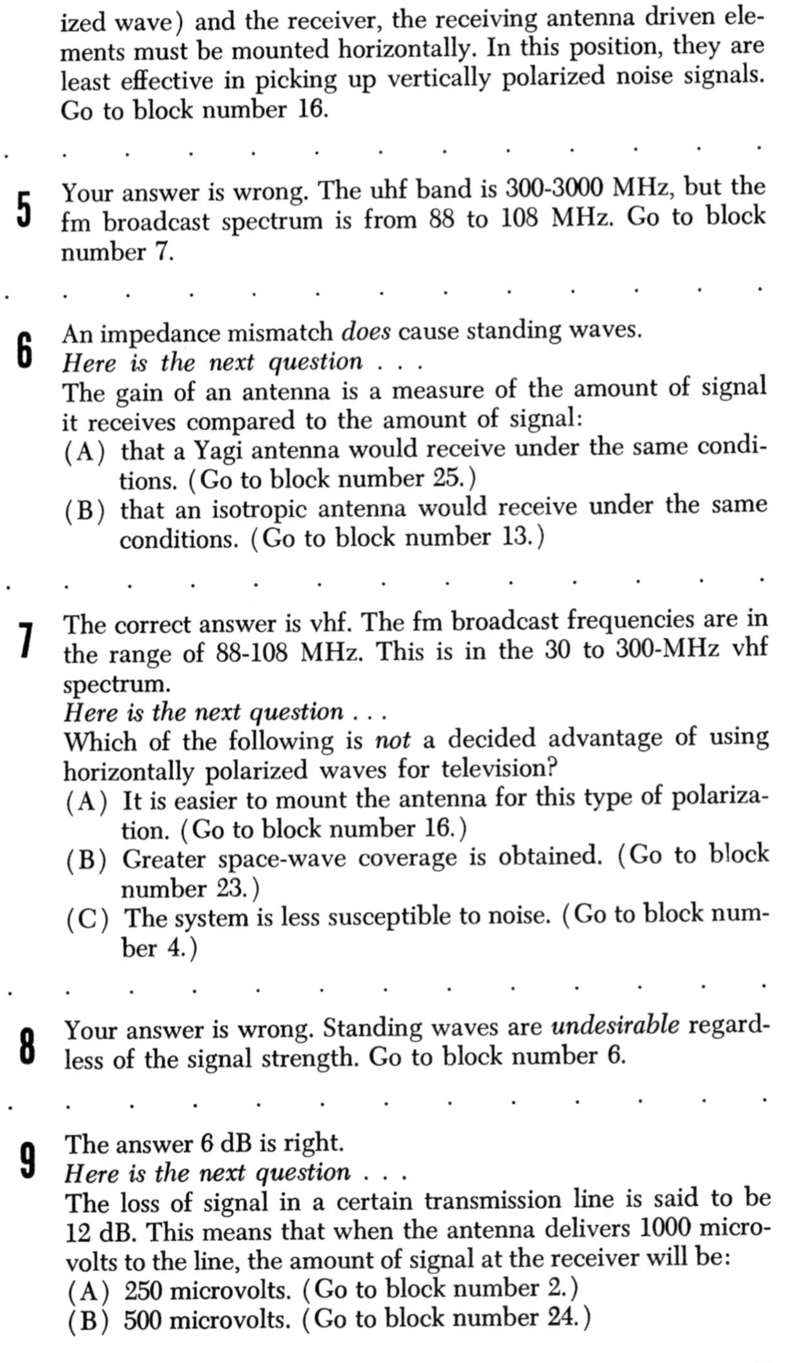

ized wave) and the receiver, the receiving antenna driven elements must be mounted horizontally. In this position, they are least effective in picking up vertically polarized noise signals. Go to block number 16.

5 Your answer is wrong. The uhf band is 300-3000 MHz, but the fm broadcast spectrum is from 88 to 108 MHz. Go to block number 7.

6 An impedance mismatch *does* cause standing waves.

Here is the next question . . .

The gain of an antenna is a measure of the amount of signal it receives compared to the amount of signal:

(A) that a Yagi antenna would receive under the same conditions. (Go to block number 25.)

(B) that an isotropic antenna would receive under the same conditions. (Go to block number 13.)

7 The correct answer is vhf. The fm broadcast frequencies are in the range of 88-108 MHz. This is in the 30 to 300-MHz vhf spectrum.

Here is the next question . . .

Which of the following is *not* a decided advantage of using horizontally polarized waves for television?

(A) It is easier to mount the antenna for this type of polarization. (Go to block number 16.)

(B) Greater space-wave coverage is obtained. (Go to block number 23.)

(C) The system is less susceptible to noise. (Go to block number 4.)

8 Your answer is wrong. Standing waves are *undesirable* regardless of the signal strength. Go to block number 6.

9 The answer 6 dB is right.

Here is the next question . . .

The loss of signal in a certain transmission line is said to be 12 dB. This means that when the antenna delivers 1000 microvolts to the line, the amount of signal at the receiver will be:

(A) 250 microvolts. (Go to block number 2.)

(B) 500 microvolts. (Go to block number 24.)

. . . .

10 The ability of an antenna to receive signals from only one direction is a measure of its horizontal directivity. This is an important factor in its ability to reject multipath (ghost) signals.
Here is the next question . . .
In an MATV system, coaxial cable is preferred over twin lead because:
(A) it is less expensive. (Go to block number 18.)
(B) it is unaffected by nearby metal objects, and it does not radiate interference. (Go to block number 21.)

. . . .

11 Your answer is wrong. Although foam-filled twin lead is better than flat twin lead, it is still susceptible to loss due to accumulations of impurities and moisture on its surface. Go to block number 26.

. . . .

12 The antennas should be as far apart as possible.
Here is the next question . . .
An example of a balanced line is:
(A) coaxial cable. (Go to block number 14.)
(B) twin lead. (Go to block number 20.)

. . . .

13 The *gain* is a measure of the ability of an antenna to capture a signal compared to the ability of an isotropic antenna to capture the same signal when mounted at the same point. (Instead of an isotropic, a dipole may be the reference antenna.)
Here is the next question . . .
The measure of the ability of an antenna to receive a signal from one horizontal direction and reject signals from other horizontal directions is its:
(A) sensitivity. (Go to block number 15.)
(B) horizontal directivity. (Go to block number 10.)

. . . .

14 Your answer is wrong. Go to block number 20.

. . . .

15 Your answer is wrong. Go to block number 10.

16 Your answer is right. A vertical dipole is no more difficult to mount than a horizontal dipole. At frequencies below Channel 2 this could be an advantage because of the required length of the antenna elements, but at television frequencies this is not a decided advantage.

Here is the next question . . .

Which of the following statements is *not* true?

(A) A disadvantage of using horizontally polarized waves for television is that the system is more susceptible to ghosts. (Go to block number 3.)

(B) An isotropic antenna is a theoretical antenna that radiates or receives equally well in all directions. (Go to block number 22.)

17 Your answer is wrong. Doubling the *voltage* corresponds to a 6-dB gain. Go to block number 9.

18 Your answer is wrong. Coaxial cable is *more* expensive than twin lead. Go to block number 21.

19 Your answer is wrong. Mounting the antennas close to each other is undesirable. Go to block number 12.

20 The correct answer is twin lead. It is said to be balanced to ground because the voltages on both leads are the same with reference to ground. Coaxial cable is unbalanced because one lead is grounded and the other lead has a signal voltage on it.

Here is the next question . . .

Fm broadcast stations are assigned frequencies in the:

(A) uhf band. (Go to block number 5.)

(B) vhf band. (Go to block number 7.)

21 Besides the two advantages given, coaxial cable is also less likely to pick up stray noise signals.

Here is the next question . . .

On an installation for a color receiver, which would be the better choice for a transmission line (expense not withstanding)?

(A) Foam-filled twin lead (Go to block number 11.)
(B) 300-ohm shielded twin lead (Go to block number 26.)

. . . .

22 Your answer is wrong. The statement is actually a definition of an isotropic antenna. Go to block number 3.

. . . .

23 Your answer is wrong. When the transmitting and receiving antennas are located several wavelengths above ground, as in the case of television, better space-wave coverage is obtained with horizontally polarized waves. Go to block number 16.

. . . .

24 Your answer is wrong. A loss of 12 dB means that the voltage is halved (to obtain −6 dB) and then halved again (to obtain −12 dB). Go to block number 2.

. . . .

25 Your answer is wrong. A Yagi has a gain of about 10 dB, which means that the Yagi receives a greater amount of signal than a theoretical isotropic antenna under the same conditions. Go to block number 13.

. . . .

26 If cost is not a factor, 300-ohm shielded twin lead would be better.

You have now completed the programmed questions and answers.

. . . .

PRACTICE TEST

1. The spectrum of frequencies between 30 and 300 MHz is classified as:

(a) shf.
(b) uhf.
(c) vhf.
(d) hf.

2. When a wave is horizontally polarized,

(a) its magnetic field is horizontal—that is, parallel to the surface of the earth.
(b) its electric field is horizontal—that is, parallel to the surface of the earth.

3. Vhf television signals are:

(a) horizontally polarized.
(b) vertically polarized.
(c) not polarized.

4. **Which of the following is *not* an advantage of using horizontally polarized waves for television?**

   (a) Less interference with vertically polarized noise signals.
   (b) Less expensive to transmit.
   (c) Less loss for the ground wave.

5. **The television signal arrives at the receiving antenna as:**

   (a) a ground wave.
   (b) a sky wave.
   (c) a space wave.

6. **Which of the following would *not* increase the line-of-sight distance of a television signal?**

   (a) Raise the height of the transmitting antenna.
   (b) Raise the height of the receiving antenna.
   (c) Increase the length of the driven element on the receiving antenna.

7. **For a line-of-sight transmitted signal, the cutoff point is not at the horizon. Actually, the signal travels:**

   (a) beyond the horizon by some 15 percent before the signal strength begins to drop rapidly.
   (b) beyond the horizon by some 50 percent before the signal strength begins to drop rapidly.

8. **The polarization of a uhf television signal is:**

   (a) unaffected by vegetation.
   (b) affected by moisture in the air.

9. **The bending of radio waves back to earth from the ionosphere would not normally occur for which of the following frequencies?**

   (a) 18 MHz.
   (b) 88 MHz.

10. **For a Yagi antenna, which of the following elements would be longest?**

    (a) A director.
    (b) A driven element.
    (c) A reflector.

11. **The wavelength of a 210-MHz signal is:**

    (a) 18.2 feet.
    (b) 16.4 feet.
    (c) 8.2 feet.
    (d) 4.68 feet.

12. **When amplitude modulation is employed, the noise susceptibility of a system:**

    (a) is greatly affected by the relative position of the moon.
    (b) depends upon the amount of moisture in the air.
    (c) varies directly as the bandwidth.

13. **The image produced by a signal that is reflected from a building will:**

    (a) appear on the screen of the receiver as a weak ghost on the left of the image produced by the direct signal.
    (b) appear on the screen of the receiver as a weak ghost on the right of the image produced by the direct signal.

14. **A *smear ghost* is produced by:**

(a) a number of ghosts that are closely spaced.
(b) a ghost caused by the signal being picked up on the transmission line.
(c) interference due to cross coupling of the receiver i-f and audio stages.

15. Maximum transfer of power between the transmitter and the receiver, when vertical polarization is employed, will occur when the receiving antenna dipoles are:

(a) mounted horizontally.
(b) mounted at an angle of 45° with the surface of the earth.
(c) mounted vertical with respect to the surface of the earth.

16. The purpose of twisting a 300-ohm twin-lead transmission line, as it is routed from the antenna to the receiver, is:

(a) to increase the line inductance.
(b) to reduce the line capacitance.
(c) to lower the line impedance.
(d) to reduce the amount of noise pickup by the line.

17. For the maximum transfer of power between the antenna and the receiver:

(a) care must be taken that the transmission line impedance does not match the antenna or receiver impedance.
(b) the transmission line impedance need only match the antenna impedance.
(c) the transmission line impedance need only match the receiver impedance.
(d) the transmission line must match both the antenna and the receiver impedances.

18. Which of the following is *not* true regarding an impedance mismatch between the transmission line and the antenna and the receiver?

(a) A certain amount of mismatch is desirable in strong signal areas.
(b) A mismatch can cause a ghost image.
(c) A mismatch is the cause of less than the maximum power being delivered to the receiver.

19. In comparing the ghost image produced by a multipath signal and by an impedance mismatch, which of the following is *not* true?

(a) To determine the cause of the ghost, the antenna can be turned to a different direction. If this affects the ghost image, then the ghost is produced by multipath distortion rather than by impedance mismatch.
(b) Ghosts cannot be caused by impedance mismatch.
(c) A ghost caused by an impedance mismatch will normally be closer to the desired image than a ghost caused by multipath signals.

20. Another name for a half-wave dipole antenna is:

(a) Marconi antenna.
(b) Zepp-fed antenna.
(c) Hertz antenna.
(d) Yagi antenna.

21. The impedance of a half-wave center-fed dipole antenna is:

(a) 72 ohms.
(b) 200 ohms.
(c) 300 ohms.
(d) infinite.

22. The physical length of an antenna at television frequencies is:

(a) about 5% of its electrical length.
(b) about 95% of its electrical length.
(c) about 5% greater than its electrical length.
(d) about 95% greater than its electrical length.

23. If an imaginary line was drawn between the transmitter and a receiving dipole, the dipole elements would be:

(a) at right angles to the line.
(b) parallel with the line.
(c) at an angle of 45° with the line.

24. A horizontal polar pattern of a simple dipole should indicate that it is:

(a) unidirectional.
(b) bidirectional.
(c) nondirectional.

25. Which of the following transmission lines would have an impedance that would be correctly matched to the impedance of a simple dipole?

(a) A 72-ohm coaxial cable.
(b) A 300-ohm twin lead.

26. An ohmmeter across the transmission line connected to a dipole antenna (not a folded dipole) should show:

(a) about zero ohms.
(b) 75 ohms.
(c) 300 ohms.
(d) infinity.

27. A folded dipole has an impedance of approximately:

(a) 72 ohms.
(b) 150 ohms.
(c) 300 ohms.
(d) 450 ohms.

28. In comparing a folded dipole with a simple dipole which of the following is *not* true?

(a) The folded dipole is more directive than the simple dipole.
(b) The folded dipole has a higher impedance than the simple dipole.
(c) The folded dipole is more rigid in structure and can withstand a higher wind pressure than can the simple dipole.

29. The reflector and the directors on a Yagi antenna are known as:

(a) driven elements.
(b) vestigial elements.
(c) parasitic elements.

30. In comparing the Yagi antenna with a simple dipole antenna, which of the following is *not* true?

(a) The Yagi will produce a greater gain.
(b) The Yagi will accept a wider band of frequencies.
(c) The Yagi will have a narrower horizontal directional pattern.

31. An ohmmeter across the transmission line connected to a folded dipole antenna should show:

(a) about zero ohms.
(b) 75 ohms.
(c) 300 ohms.
(d) infinity.

32. For a stacked array, the antennas are usually spaced:

(a) a wavelength apart.
(b) a half-wavelength apart.
(c) a quarter wavelength apart.
(d) any spacing that is convenient.

33. **Which of the following is *not* an advantage of stacking antennas?**

(a) An increased gain.
(b) A narrower vertical directivity which makes it possible for the antenna to discriminate better against reflected ground waves.
(c) An increase in impedance which reduces the possibility of standing waves on the transmission line.

34. **Antennas with corner reflectors are seldom used for vhf reception because:**

(a) their impedance is too high.
(b) their gain is too low.
(c) they are much too bulky and heavy.

35. **There are many types of guy wires in use. A guy wire rated 8 × 10 would be:**

(a) 8′ long and 10″ wide.
(b) number 10 wire with 8 strands.
(c) number 8 wire with 10 strands.
(d) none of these.

36. **In reference to a flat ribbon parallel line, which of the following is correct?**

(a) It is designed primarily for outside use.
(b) It has a greater loss than coaxial cable.
(c) Attenuation losses in this type of line are increased when moisture collects on the surface.

37. **Which of the following is *not* a factor in determining the characteristic impedance of a twin-lead transmission line?**

(a) Length of line.
(b) The distance between wires.
(c) The type of material between wires.
(d) Diameter of wires used for conductors.

38. **In order to match a balanced transmission line to a coaxial cable, you would choose a:**

(a) multiset coupler.
(b) balun.
(c) bifolar transformer.

39. **The service area of a television transmitter for rural and residential areas is normally defined by a contour line beyond which the signal strength falls below about:**

(a) 2000 microvolts per meter.
(b) 1500 microvolts per meter.
(c) 500 microvolts per meter.
(d) 1000 microvolts per meter.

40. **A television preamplifier is mounted on an antenna. How does this preamplifier normally get its dc operating power?**

(a) It is battery powered.
(b) With solar batteries.
(c) Through a separate line leading to the power supply in the receiver.
(d) Through the same transmission line that delivers the signal to the set.

41. **Which of the following antennas could be designed so that it is capable of receiving the entire range of TV frequencies from Channel 2 to Channel 83?**

(b) A conical antenna.
(c) A folded-dipole stacked array.
(d) A single dipole with one director.
(e) A log-periodic antenna.

42. If the antenna is properly oriented, the director of a Yagi is:

(a) closest to the station.
(b) furthest from the station.
(c) further from the station than the driven element, but closer to the station than the reflector.
(d) none of these.

43. For MATV systems, a 1000-microvolt signal may also be referred to as:

(a) zero dBmV.
(b) 100 millivolts.
(c) 3 dB.
(d) 6 dB.

44. The trouble symptom is white flashes and static in the picture. Which of the following is a likely cause?

(a) There is impedance mismatch between the antenna and transmission line and the receiver.
(b) There is an impedance mismatch between the transmission line and the receiver.
(c) There is a broken wire in the transmission line.
(d) There is a poor ground on the lightning arrestor.

45. The receiving antenna in a radio a-m broadcast receiver is usually:

(a) a rhombic antenna.
(b) a ferrite antenna.
(c) a balun.
(d) a Zepp antenna.

46. In an area with a reflected-signal problem, which of the following antennas will give the best results?

(a) Rabbit ears.
(b) A simple dipole with reflector.
(c) A conical with reflector.
(d) A Yagi.

47. In a certain antenna preamplifier, the output power is twice the input power. This amplifier has:

(a) a 6-dB gain.
(b) a 3-dB gain.
(c) zero dBmV.
(d) a loss of 6 dB.

48. A negative dB gain indicates that there is:

(a) a negative resistance in the circuit.
(b) an impossible situation.
(c) a loss in the circuit.
(d) an improvement in signal strength.

49. In comparing the operation of two television transmitters, one broadcasting in the vhf range and the other broadcasting in the uhf range, which of the following is true?

(a) The uhf station must transmit more power in order to get the same amount of coverage.
(b) The vhf station must transmit more power to get the same amount of coverage.

50. When a preamplifier is used for improving the signal-to-noise ratio at a television installation:

(a) the preamplifier could be mounted as close to the antenna as possible.
(b) the preamplifier should be mounted as close to the receiver as possible.

4

Electronic Components

KEYED STUDY ASSIGNMENTS

Howard W. Sams Photofact Television Course
Chapter 1—Cathode-Ray Tube: Beam Formation and Electrostatic Control
Chapter 2—Cathode-Ray Tube: Electromagnetic Control of the Beam

Howard W. Sams Color-TV Training Manual
Chapter 9—The Color Picture Tube and Associated Circuits

This assignment covers the theory of monochrome and color picture tubes, and also covers the theory of their related components such as focus coils and yoke assemblies.

In this chapter, we will review some of the components used in home-entertainment equipment.

IMPORTANT CONCEPTS

Resistors

All resisors can be classified as being either *linear* or *nonlinear.* When the voltage across a linear resistor is doubled, the current through it is also doubled—provided that the resistor is operated within its power rating. Another way of saying this is that linear resistors follow Ohm's law when they are operated within their temperature limits.

A nonlinear resistor has a wide variation in resistance for a small change in temperature. (The change in temperature can be caused by a change in current through the resistor.)

Linear Fixed Resistors—There are three kinds of linear, fixed-value resistors in popular use: *carbon-composition, wirewound,* and *film resistors.* The first two have been in use for a long time. The film resistors are made by depositing a resistive film on a ceramic or other suitable base, attaching leads, and then coating the assembled component with an insulated coating.

Resistors have a *temperature coefficient* (*TC*) which is an indication of the amount of resistance change that will occur for a given change in temperature. For example, carbon-composition resistors have a *positive temperature coefficient,* which means that their resistance *increases* with an increase in temperature. *Thermistors,* which are nonlinear resistors, usually have a *negative temperature coefficient,* which means that their resistance *decreases* with an increase in temperature.

It is presumed that you are familiar with the resistor color code including the tolerance band. When the fourth band is gold, the tolerance is ± 5 percent; when the fourth band is silver, the tolerance is ± 10 percent; and, when there is no color for the fourth band the tolerance is ± 20 percent.

When the third band is gold, it is a *multiplier* and its value is 0.1. For example, a resistor that is color coded yellow, violet, and gold has a resistance value of 47 × 0.1 or 4.7 ohms. When the third band is colored silver, it is a *multiplier,* and its value is 0.01. Thus, a resistor that is color coded yellow, violet, and silver has a resistance value of 47 × 0.01 or 0.47 ohms.

You may see a resistor with a wide band and a narrow band. This type of color coding always indicates that the resistance value is less than ten. A color code of yellow (wide) and violet (narrow) means a resistance value of 4.7 ohms.

Remember that black in the third band means *no multiplier,* and the value of resistance is between 10 and 100 ohms. Another thing to remember is that resistors with radial leads are sometimes called *BED* resistors because their color code is read in the following order: *B*ody *E*nd *D*ot.

Nonlinear Fixed Resistors—A number of different types of nonlinear resistors are being used in electronic equipment. *Thyrite* resistors are made of silicon carbide. *Thyrite* is a trade name for these components which were first introduced by the General Electric Co. They are characterized by a highly negative temperature coefficient.

Thermistors are nonlinear resistors which normally have a negative temperature coefficient. However, some of these, called *positive TC* thermistors, increase in resistance when their temperature rises. Thermistors are made of a semiconductor material that is doped to get a desired resistivity.

Varistors, which are also called *voltage-dependent resistors* or *VDR's,* exhibit a relatively large amount of resistance change with a small change in voltage across their terminals. When the voltage across a VDR increases, its resistance decreases.

A *photosensitive resistor,* or, *Light Dependent Resistor* (*LDR*), has a resistance value that depends upon the amount of light falling on it. These components have been used in automatic brightness and contrast circuits to compensate for changes in room brightness.

Variable Resistors—Variable resistors are classified according to their *taper.* The taper refers to the way the resistance changes with rotation of the shaft. A variable resistor with a *linear* taper varies in resistance directly with the amount of rotation of the shaft. However, the resistance of a *nonlinear* variable resistor does not vary directly with the amount of shaft rotation. A linear variable resistor is NOT suitable for a volume control or brightness control because the ears and eyes of humans are nonlinear. By using a resistor with a nonlinear taper, it appears to our ears that the amount of change in sound volume is in direct proportion to the amount of shaft rotation.

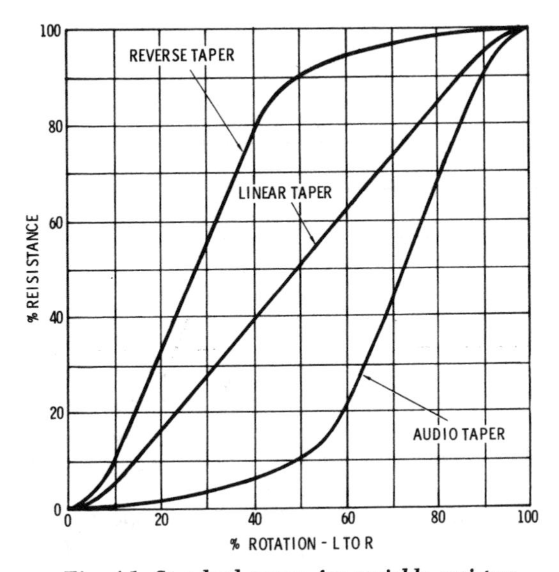

Fig. 4-1. Standard tapers for variable resistors.

Fig. 4-1 shows a graph of the three types of tapers available for variable resistors. Note that fifty percent of the maximum resistance is obtained for fifty percent of the shaft rotation in the case of a linear taper. For an audio taper, 10 percent of the maximum resistance is obtained for 50 percent of the shaft rotation, while 90 percent of the maximum resistance is obtained by rotating the shaft of a reverse-taper variable resistor halfway.

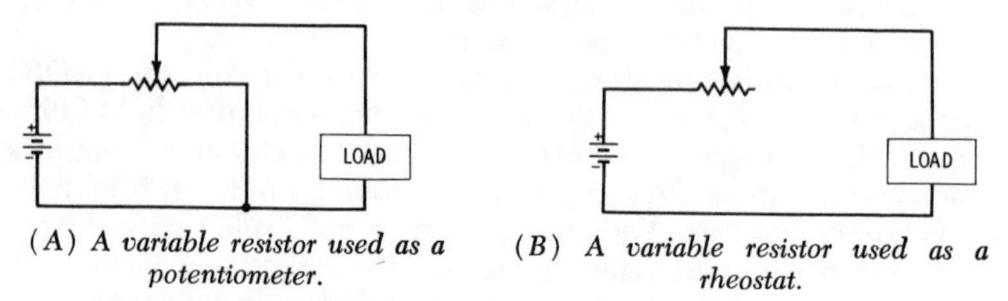

(A) *A variable resistor used as a potentiometer.*

(B) *A variable resistor used as a rheostat.*

Fig. 4-2. Two ways of connecting a variable resistor in a circuit.

Although variable resistors are often called *pots* (for potentiometers), there are actually two ways to connect a variable resistor in a circuit. When they are used for varying voltage, they are known as potentiometers; and, when they are used for varying current, they are known as rheostats. Fig. 4-2 shows the two connections.

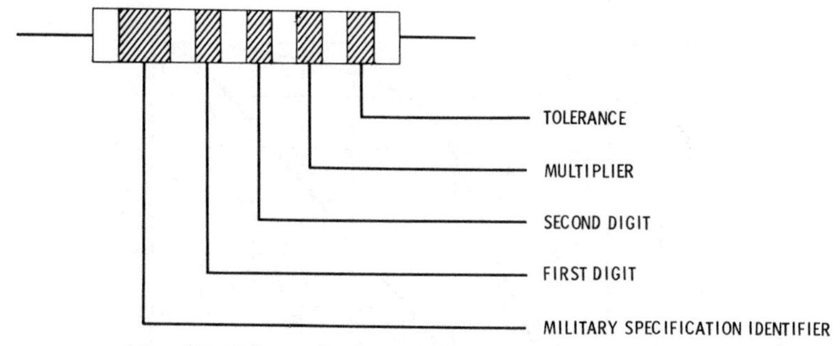

Fig. 4-3. Color code for tubular encapsulated resistors.

Inductors

Tubular encapsulated rf chokes are color coded with a five-band system like the one shown in Fig. 4-3. Note that the fourth band is the multiplier. If the fourth band is gold or silver, then the value of inductance is less than ten microhenrys.

Figs. 4-4, 4-5, and 4-6 show the color codes for power transformers, audio-frequency transformers, and i-f transformers.

Toroidal inductors are coils that are wound on a doughnut-shaped (toroidal) core. They are more expensive than other types of coils, but they have advantages that justify the higher cost in certain applications. They have a higher inductance for a given size of inductor, they are self-shielding, and they are highly stable.

A *ferrite bead* is a small bead of ferrite (that is, iron oxide) that acts as an rf choke. It has certain advantages over the conventional

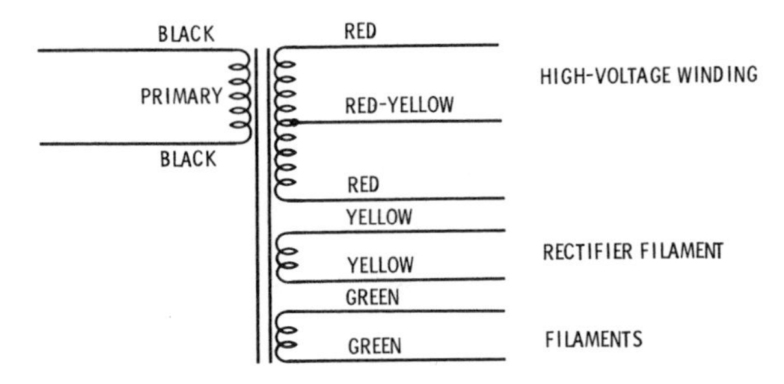

Fig. 4-4. Color code for power transformers.

wirewound type of choke. While wirewound chokes have dc resistance due to the wire resistance, ferrite beads do not. Also, ferrite beads have a smaller physical size for a given amount of inductance, and they do not have the distributed capacitance associated with wire-wound types. This means that they do not tend to peak at some frequency.

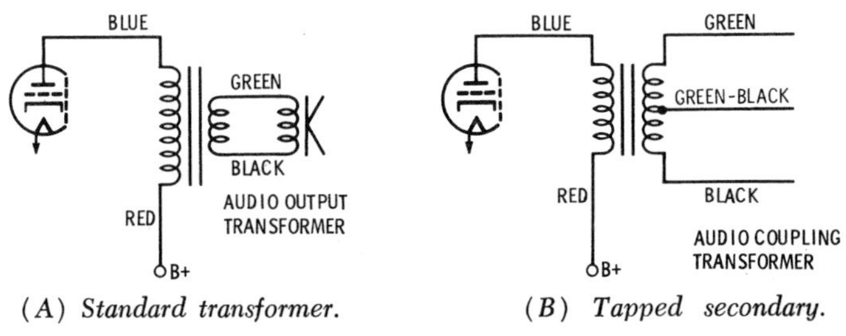

(*A*) *Standard transformer.* (*B*) *Tapped secondary.*

Fig. 4-5. Color code for audio transformers.

Variable inductors are used for tuning automobile receivers. This is called *permeability tuning*. The inductors used in this application usually have variable-pitch windings, which means that the windings are crowded at one end and spread out at the other end. A nonlinear winding is necessary in order to get a linear (or near-linear) tuning dial for the receiver.

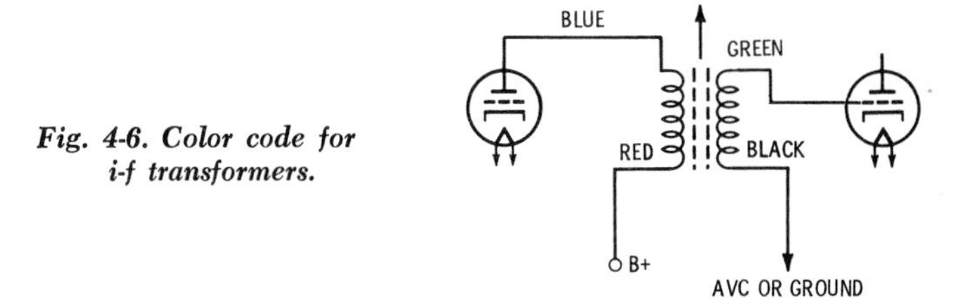

Fig. 4-6. Color code for i-f transformers.

Capacitors

Capacitors, like resistors and inductors, may be either *fixed* or *variable.* Fixed capacitors are identified a number of different ways. They are *electrolytic* if the dielectric material is an electrolyte, or *electrostatic* if the dieletric is not an electrolyte. Electrostatic capacitors are usually identified by the type of material used for a dielectric. Examples are paper and mica. Some of the more popular capacitors will be discussed briefly.

Aluminum Electrolytic Capacitors—Aluminum electrolytic capacitors use an aluminum-oxide coating on the surface of an aluminum foil for a dielectric. The foil is usually etched to obtain a larger surface area. Fig. 4-7 shows this construction. Because the oxide coating is so thin, the capacitance is large. (Remember that the capacitance increases with an increase in the area of plates and increases with a decrease in the distance between the plates.) Electrolytic capacitors may be *polarized,* which means that they have a positive and a negative terminal and must not be connected across an ac voltage. Some types of electrolytic capacitors are nonpolarized, which means that they can be subjected to an ac voltage for a short period of time. Although the capacitance of aluminum electrolytics is high, their leakage resistance is low. Also, they tend to deteriorate with age.

Ceramic Capacitors—Ceramic capacitors are made by depositing a metal coating on a ceramic tube or disc. Fig. 4-8 shows a typical construction. They have the highest capacitance, for a given size, of the smaller capacitance types. Values range from 0.5 pF to 0.05 μF, with breakdown voltages as high as 3 kilovolts. An important feature is their *temperature coefficients.* They can be made with *positive temperature coefficients* (meaning that their capacitance increases with temperature rise), or *negative temperature coefficients* (meaning that their capacitance decreases with temperature rise).

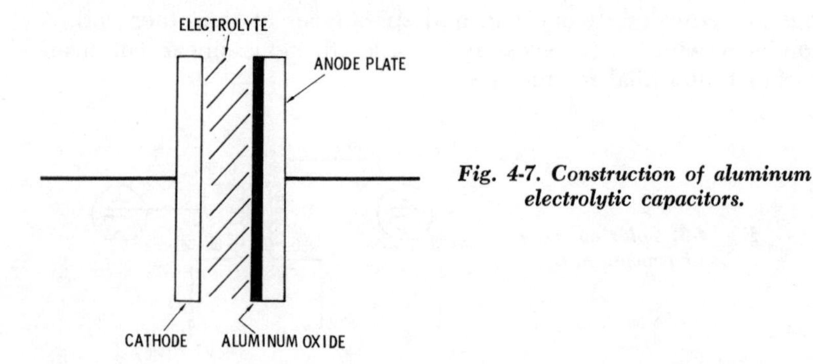

Fig. 4-7. Construction of aluminum electrolytic capacitors.

Ceramic capacitors with a zero temperature coefficient, meaning that their capacitance does not change with temperature, are available. More will be said about temperature coefficients later.

Glass Capacitors—Glass capacitors use a thin glass wafer for a dielectric. The capacitor plates are made by depositing a metal film on both sides of the glass. The complete assembly is molded in glass. Typical capacitance values range from 0.5 pF to 0.01 μF with voltages to 500 volts. They have excellent temperature stability and a very high dielectric leakage resistance. However, they are costly and subject to breakage.

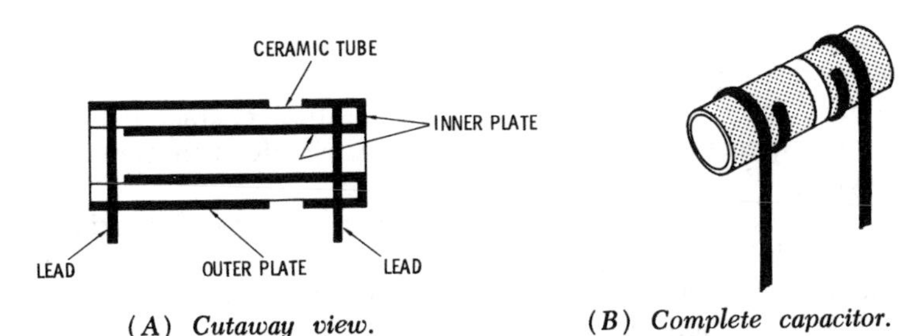

(*A*) *Cutaway view.* (*B*) *Complete capacitor.*

Fig. 4-8. Construction of a ceramic capacitor.

Metallized Paper or Metallized Plastic Capacitors—Metallized paper or metallized plastic capacitors are made by depositing a metal film on opposite sides of the paper or plastic dielectric. Two important characteristics of this type are: *self healing* and *higher losses.* Values from 0.5 pF to 0.1 *μF* and breakdown-voltage ratings to 0.6 kilovolts are common.

Mica Capacitors—Mica capacitors employ thin sheets of mica for a dielectric. Metal foil may be used for the plates, or the mica may be metallized—that is, the metal may be deposited directly on the mica. Silver mica capacitors are an example of metallized mica capacitors. They have a high breakdown-voltage rating, low loss, but usually are not made with large capacitance values. (Values usually range from 5 pF to 0.1 μF with voltage breakdown values up to 2500 volts.) Silver mica capacitors are used in critical circuits, especially at high rf frequencies, because of their low inductance, temperature stability, and small size.

Paper Capacitors—Paper capacitors use paper for their dielectric. With impregnated paper capacitors, the paper is impregnated with oil or wax. They are usually made in values ranging from 1000 pF to 20 μF, and their breakdown voltage may be as high as two or three kilovolts. Their voltage ratings, capacitance values, and dielectric losses increase with temperature. The operating temperature of a paper capacitor is increased by combining the paper with a plastic film. This also increases the expected life of the capacitor, but at the same time it increases the cost.

Plastic Film Capacitors—Plastic film capacitors have a plastic film as their dielectric. They are characterized by high insulation resistance, good stability, and low losses. *Mylar,* cellophane, and *Teflon* are examples of plastic films used. They are made with capacitance values ranging from 1000 pF to 1 μF, and breakdown voltages as high as two kilovolts.

Tantalum Electrolytic Capacitors—Tantalum electrolytic capacitors employ tantalum oxide for a dielectric. Since tantalum oxide has nearly twice the dielectric constant of aluminum oxide, and is very stable under varying temperatures, the advantages of capacitors using this type of dielectric are apparent. They have a higher capacitance for a given size, better temperature characteristics, a longer life expectancy, and a much better shelf life than aluminum electrolytics.

Variable Capacitors—There are a number of ways of making a capacitor variable. You can vary the area of the facing plates, vary the distance between the plates, or vary the type of dielectric used. The first two methods are the ones usually employed. Variable capacitors used for tuning radios are nonlinear. This means that their capacitance does not vary directly with the amount of shaft rotation. They must be made nonlinear in order to get a linear (or near-linear) dial.

Capacitor Ratings—Capacitors are placed in parallel to obtain a higher capacitance value and in series to obtain a higher working voltage. As a general rule, using a capacitor with a higher voltage rating than required will increase its life expectancy. However, in the case of electrolytic capacitors, not much is gained by increasing the working voltage rating by more than 20 or 30 percent above the required value.

The equivalent series resistance (ESR) of a capacitor is a measure of how much opposition it will offer to the flow of dc. As the temperature of an electrolytic capacitor goes up, its ESR goes down, but for other types of capacitors the ESR may rise with temperature.

The dissipation factor of a capacitor is the ratio of its ESR to its capacitive reactance. It is usually expressed as a percentage value. The reciprocal of the dissipation factor is the Q. The power factor of a capacitor is the ratio of its ESR to the total impedance of the capacitor.

The insulation resistance of a capacitor tells the effective amount of resistance bypassing the capacitor. The lower the insulation resistance, the more quickly a capacitor will discharge when the charging voltage is removed. A high insulation resistance is desirable.

The *voltage rating* of a capacitor tells the amount of dc voltage that can be placed across the capacitor leads continuously without danger of the insulation breaking down. This is sometimes called the dc working voltage (Vdcw). The Vdcw rating of a capacitor depends upon the type and the thickness of the dielectric. There is no positive way to convert a Vdcw rating to an ac voltage rating. However, the Vdcw rating should at least be greater than the peak value of ac voltage across the capacitor.

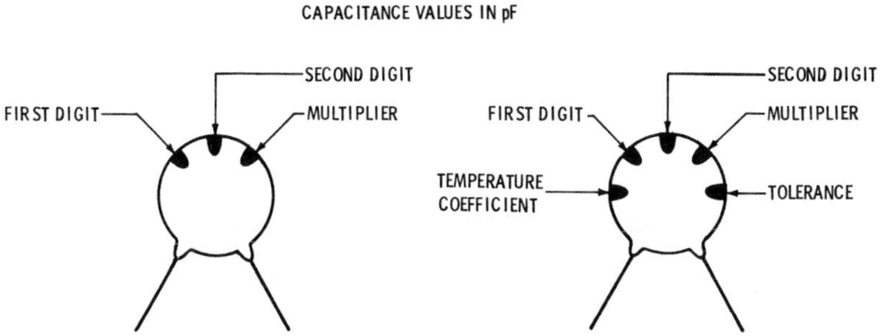

CAPACITANCE VALUES IN pF

FIRST DIGIT
SECOND DIGIT
MULTIPLIER

(*A*) *3-dot system.*

FIRST DIGIT
SECOND DIGIT
MULTIPLIER
TEMPERATURE COEFFICIENT
TOLERANCE

(*B*) *5-dot system.*

TEMPERATURE COEFFICIENTS			
BLACK	NPO	BLUE	N470
BROWN	N033	VIOLET	N750
RED	N075	GREY	P030
ORANGE	N150	WHITE	GENERAL PURPOSE
YELLOW	N220	SILVER	GENERAL PURPOSE
GREEN	N330	GOLD	P100

(*C*) *Temperature coefficients.*

Fig. 4-9. Color codes for disc ceramic capacitors.

Because of the wide tolerance of electrolytic capacitors, they may be sold with a *guaranteed minimum value* (GMV) rather than by a tolerance.

The *temperature coefficient* of a capacitor is a measure of how much capacitance change will occur for a given change in temperature. If the temperature coefficient is rated at −470, or N470, it means that its capacitance will *decrease* 470 parts per million for an increase in temperature of 1°C. A rating of 0, or NP0, means that the capacitance will not change with changes in temperature. The initials NP0 mean negative positive zero.

Figs. 4-9 to 4-12 show the color code of some of the popular fixed capacitors.

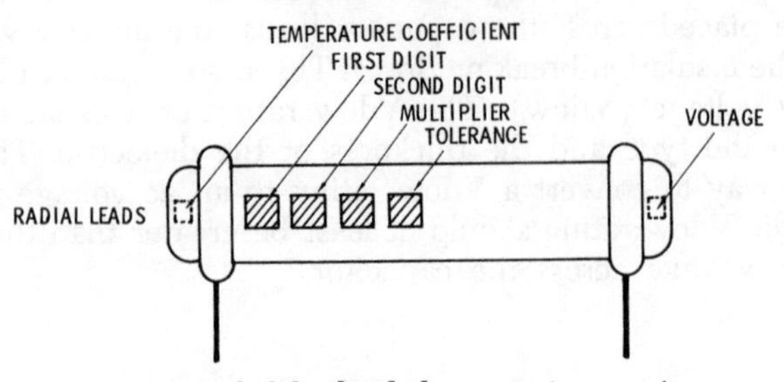

(A) Radial lead tubular ceramic capacitors.

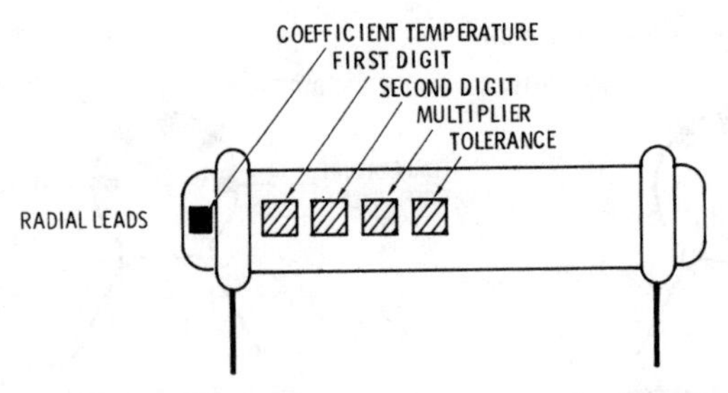

(B) Extended-range temperature-compensated tubular ceramic capacitors.

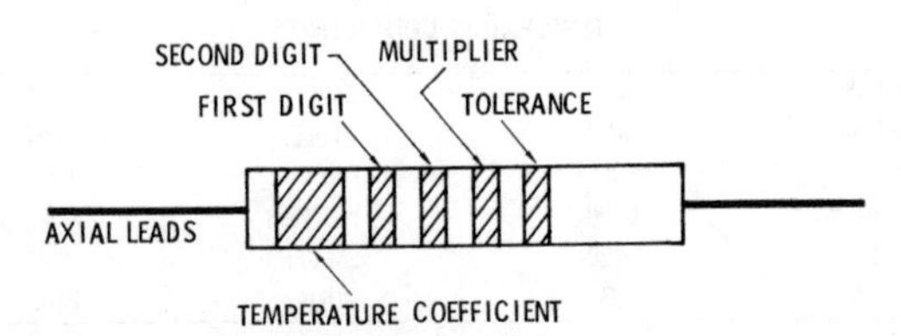

(C) Axial-lead tubular ceramic capacitors.

Fig. 4-10. Color codes for tubular ceramic capacitors.

PROGRAMMED QUESTIONS AND ANSWERS

Starting with question number 1, select the answer that you feel is correct. If you feel that (A) is correct, proceed to block number 17 as directed. If you feel that (B) is correct, proceed to block number 9 as directed. If you feel that more than one answer is correct, choose the one that you think is the *most* correct.

. . . .

1 When the current through a certain thermistor is 10 mA, the voltage across it is 10 volts. When the current is changed to 5 mA, the voltage across the thermistor:

(A) cannot be determined from the information given. (Go to block number 17.)

(B) is five volts. (Go to block number 9.)

. . . .

2 Your answer is wrong. A Thyrite resistor has a highly negative temperature coefficient. Go to block number 12.

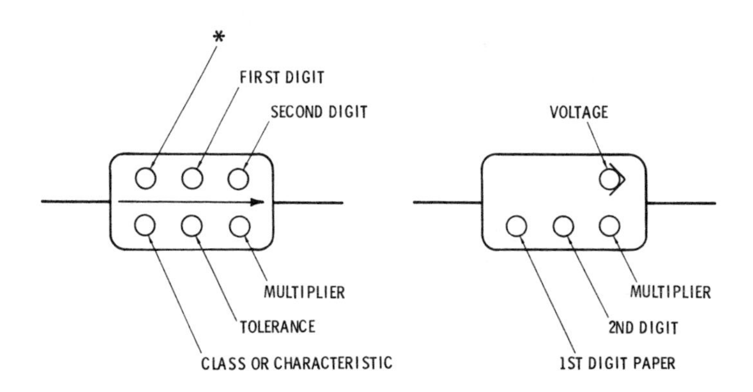

(A) *EIA and military code for paper or mica capacitors.*

(B) *Commercial code for paper capacitors.*

Fig. 4-11. Color codes for flat molded capacitors.

. . . .

3 The fact that ferrite beads do not have distributed capacitance is important because it means that they will not have some undesired resonant frequency.

Here is the next question . . .

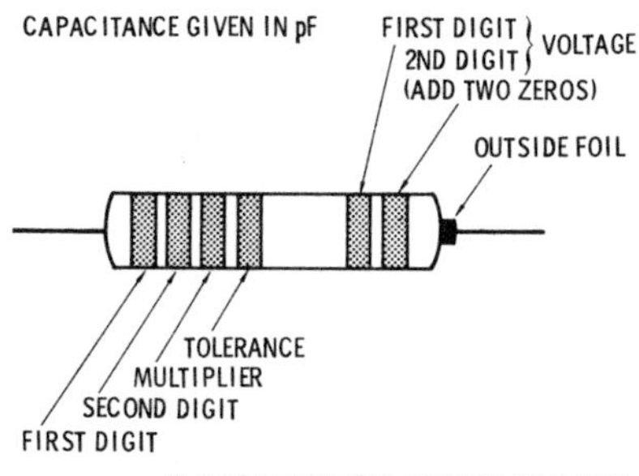

Fig. 4-12. Color code for molded paper capacitors.

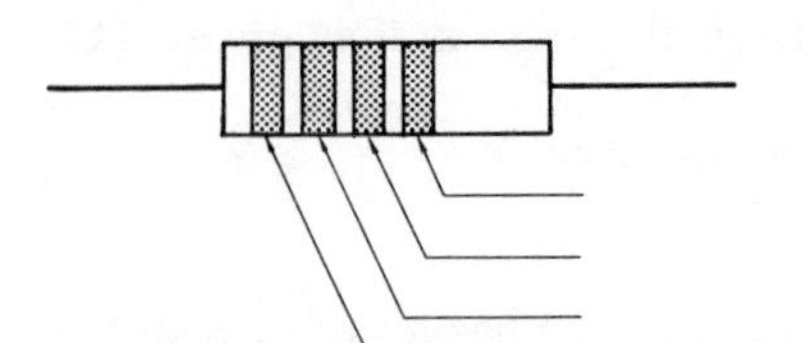

Fig. 4-13. Color-coded resistor.

For Fig. 4-13 show the colors for each band, assuming that it is a 47K resistor with a tolerance of ±5 percent. (Go to block number 7.)

. . . .

4 Another name for a voltage-dependent resistor (VDR) is varistor.

Here is the next question . . .

A designer has carefully designed an audio amplifier that is very linear. For a volume control he should choose a variable resistor with:

(A) a linear taper. (Go to block number 6.)

(B) a nonlinear taper. (Go to block number 19.)

. . . .

5 Your answer is wrong. Go to block number 3.

. . . .

6 Your answer is wrong. A volume control with a linear taper would produce a nonlinear change in volume (as heard by the ear) when the shaft is rotated. This is because the ear is nonlinear. Go to block number 19.

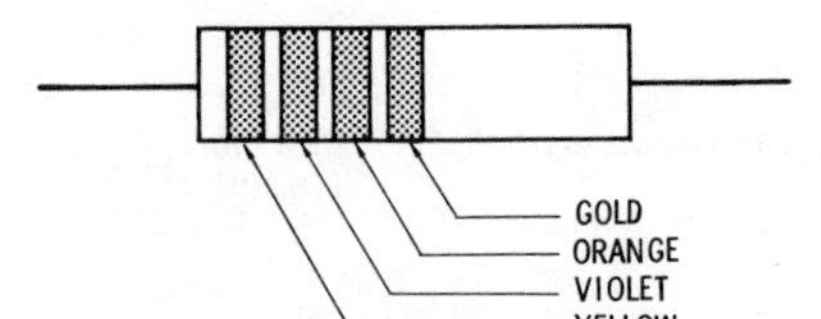

Fig. 4-14. Resistor having a resistance of 47K and a tolerance of ±5%.

. . . .

7 Your color code should look like Fig. 4-14.

Here is the next question . . .

For RC coupling between transistor audio amplifiers, which of the following types of capacitors is most likely to be used?

(A) Silver mica. (Go to block number 22.)
(B) Glass. (Go to block number 21.)
(C) Tantalum. (Go to block number 14.)

. . . .

8 The variable resistor has a linear taper.
Here is the next question . . .
Which of the following is an advantage of a ferrite bead over a wirewound choke or coil?
(A) Ferrite beads have a much larger resistance to the flow of dc current. (Go to block number 5.)
(B) Ferrite beads do not have a distributed capacitance. (Go to block number 3.)

. . . .

9 Your answer is wrong. A thermistor is a nonlinear resistor. If you change the current through it, you change its temperature. This, in turn, results in a resistance change. You cannot calculate the new voltage drop by Ohm's law or by using linear ratios. Go to block number 17.

. . . .

10 Your answer is wrong. Go to block number 8.

. . . .

11 Carbon composition resistors have a positive temperature coefficient.
Here is the next question . . .
A certain resistor is color coded as follows: brown, black, black. This resistor has a resistance value of:
(A) 10 ohms. (Go to block number 24.)
(B) 100 ohms. (Go to block number 25.)

. . . .

12 Wirewound resistors are linear.
Here is the next question . . .
Another name for VDR is:
(A) varistor. (Go to block number 4.)
(B) thermistor. (Go to block number 20.)

. . . .

13 Your answer is wrong. When gold is in the third band, the multiplier is 0.1. Go to block number 23.

. . . .

14 Tantalum capacitors can be obtained with large capacitance values. They are ideal for coupling transistor audio amplifiers.

Here is the next question . . .

A certain capacitor has a decrease in capacitance of 470 parts per million when the temperature rises 1°C. How is the temperature coefficient of this capacitor expressed? (Go to block number 26.)

. . . .

15 Your answer is wrong. Ferrite beads are small components that act like inductors. Go to block number 16.

. . . .

16 The correct answer is film resistors.

Here is the next question . . .

Carbon composition resistors have:

(A) positive temperature coefficients. (Go to block number 11.)

(B) negative temperature coefficients. (Go to block number 18.)

. . . .

17 The voltage drop cannot be calculated without additional information because the resistor is nonlinear.

Here is the next question . . .

There are three kinds of resistors in popular use. They are: carbon composition, wirewound, and

(A) ferrite bead resistors. (Go to block number 15.)

(B) film resistors. (Go to block number 16.)

. . . .

18 Your answer is wrong. It is true that a carbon-composition resistor which has been seriously overheated may have a low resistance. However, when they are operated within their normal range of temperature, they have a positive temperature coefficient. Go to block number 11.

. . . .

19 The designer wants a nonlinear taper. In fact, he wants a variable resistor with an audio taper.

Here is the next question . . .

You measure the total resistance of a variable resistor and find that it is 1000 ohms. When you turn the shaft halfway through its rotation and measure the resistance between the arm and one end, you get 500 ohms. This variable resistor has:
(A) a linear taper. (Go to block number 8.)
(B) a reverse taper. (Go to block number 10.)

. . . .

20 Your answer is wrong. A thermistor is a temperature-sensitive resistor. A VDR is a resistor that has a resistance value which is dependent upon the voltage across it. Go to block number 4.

. . . .

21 Your answer is wrong. It is desirable to use a large coupling capacitor to obtain a good low-frequency response. As you know, glass capacitors are not normally made with large capacitance values. Go to block number 14.

. . . .

22 Your answer is wrong. It is desirable to use a large coupling capacitor to obtain a good low-frequency response. As you know, silver-mica capacitors are not made with large capacitance values. Go to block number 14.

. . . .

23 The resistance value is 1.0 ohms.
Here is the next question . . .
Which of the following is an example of a linear resistor?
(A) A Thyrite resistor (Go to block number 2.)
(B) A wirewound resistor (Go to block number 12.)

. . . .

24 The resistance value is 10 ohms.
Here is the next question . . .
A resistor is color coded as follows: brown, black, gold. Its resistance value is:
(A) One ohm. (Go to block number 23.)
(B) 10 ohms. (Go to block number 13.)

. . . .

25 Your answer is wrong. The color black in the third band means that the first two digits (one and zero) are to have no multiplier. Go to block number 24.

. . . .

26 The temperature coefficient is expressed as −470 or N470. You have now completed the programmed questions and answers.

. . . .

PRACTICE TEST

1. In order to be able to calculate the amount of current through a resistor by Ohm's law, it is necessary that the resistor be:

(a) nonlinear.
(b) linear.
(c) a thermistor type.
(d) a VDR.

2. A thermistor has a wide variation in resistance for a small change in:

(a) humidity.
(b) temperature.
(c) barometric pressure.
(d) time.

3. Which of the following is NOT a linear resistor?

(a) Wirewound.
(b) Carbon composition.
(c) VDR.
(d) Film.

4. A certain type of resistor is made by depositing a resistive coating on a ceramic base, attaching leads, and coating the assembly. This type of resistor is:

(a) wirewound.
(b) carbon composition.
(c) VDR.
(d) film.

5. Which of the following statements is correct?

(a) When an increase in temperature produces an increase in resistance, the temperature coefficient is positive.
(b) When an increase in temperature produces a decrease in resistance the temperature coefficient is positive.

6. A ten-ohm resistor would be color coded:

(a) brown, black, silver.
(b) brown, black, gold.
(c) brown, black, black.
(d) brown, black, brown.

7. A resistor with radial leads is colored as follows: the end is red, the body is brown, and there is a green dot in the center. The resistance of this resistor is:

(a) 1.2 megohms.
(b) 520 ohms.
(c) 1500 ohms.
(d) 250 ohms.

8. In order to show that a resistor has a tolerance of ±10 percent:

(a) the third band is silver.
(b) the third band is gold.
(c) there is no color in the fourth band.
(d) the fourth band is silver.

9. If you increase the current through a Thyrite resistor,

(a) there will be no change in its resistance.
(b) its resistance will decrease.
(c) it will crack unless the current is increased very slowly.
(d) its resistance will increase.

10. A Thyrite resistor is placed in parallel with a thermistor. An increase in current through the parallel combination will:

(a) produce no change in the parallel resistance of the two components.
(b) cause a rapid increase in the parallel resistance of the two components.
(c) cause a slow increase in the parallel resistance of the two components.
(d) cause a decrease in the parallel resistance of the two components.

11. A photosensitive resistor is called:

(a) an LDR.
(b) an LSR.
(c) a PSR.
(d) a varistor.

12. A type of resistor that changes resistance in accordance with the amount of voltage across its terminals is called:

(a) an EDR.
(b) an ESR.
(c) a PSR.
(d) a VDR.

13. Which of the following is not a common taper for variable resistors?

(a) Sinusoidal.
(b) Linear.
(c) Audio.
(d) Reverse.

14. Assuming that the shaft is at midposition for both resistance measurements in Fig. 4-15, you can calculate that this variable resistor:

(a) is defective.
(b) has a linear taper.
(c) has a reverse taper.
(d) has an audio taper.

15. In order to determine that a resistor has a linear taper,

(a) determine if it is wire wound.
(b) look for a code letter L on the case.
(c) look for a green dot near the center terminal.
(d) make a measurement.

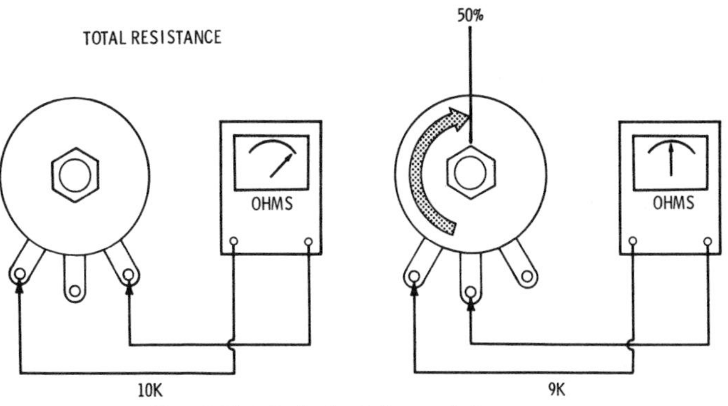

Fig. 4-15. Method of determining taper.

16. In reference to Fig. 4-16, which drawing shows the normal connection for using R_1 as a rheostat to control current through R_2?

(a) A.
(b) B.
(c) C.
(d) D.

17. In reference to Fig. 4-16, which drawing shows the normal connection for using R_1 as a potentiometer to control the voltage across R_2?

(a) A.
(b) B.
(c) C.
(d) D.

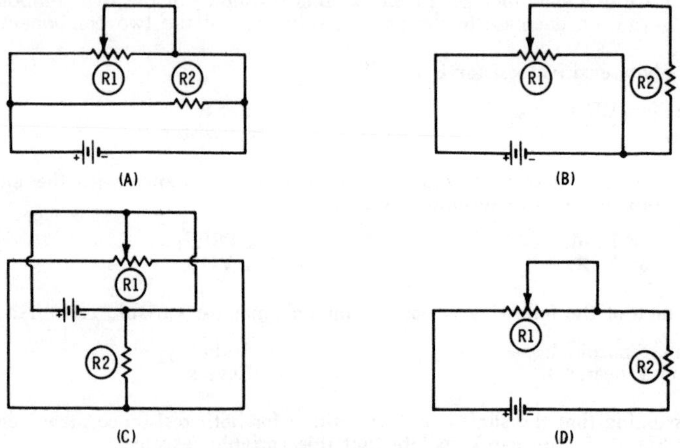

Fig. 4-16. Variable resistor circuits.

18. The B+ lead of an i-f transformer is colored:

(a) blue.
(b) red.
(c) green.
(d) black.

19. B+ lead of an audio transformer is colored:

(a) blue.
(b) red.
(c) green.
(d) black.

20. For a power transformer, the green leads are for the:

(a) primary.
(b) high-voltage plate.
(c) filament.
(d) center taps.

21. For the tubular encapsulated r-f choke coil of Fig. 4-17, which band is for the multiplier?

(a) B.
(b) C.
(c) D.
(d) A.

22. For the tubular encapsulated r-f choke coil of Fig. 4-17, which band is for the tolerance?

(a) E.
(b) D.
(c) C.
(d) A.

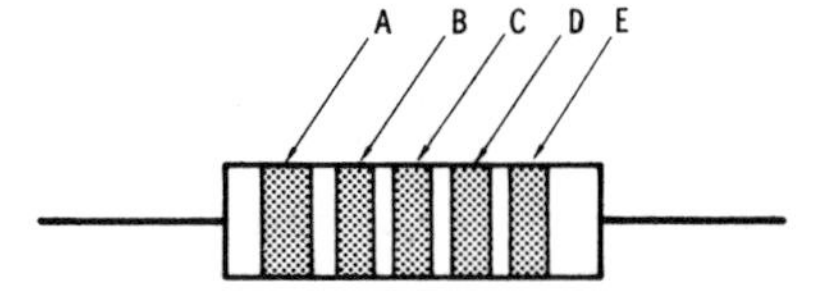

Fig. 4-17. A tubular encapsulated coil.

23. Which of the following is not an advantage of toroidal coils over other types of coils?

(a) They are less expensive.
(b) They are self-shielding.
(c) They have a higher inductance for a given size.
(d) They are highly stable.

24. A ferrite bead acts like:

(a) a resistor.
(b) an inductor.
(c) a capacitor.

25. The variable inductors used for permeability tuning in auto radios are:

(a) wound linearly in order to get a linear dial.
(b) wound nonlinearly in order to get a linear dial.

26. Which of the following is not a method commonly used for varying the capacitance of a variable capacitor?

(a) Change the type of material used for a dielectric.
(b) Change the area of plates facing each other.

27. To increase the capacitance of a capacitor:

(a) increase the distance between the plates.
(b) decrease the distance between the plates.
(c) use a dielectric material with a lower dielectric constant.
(d) reduce the area of plates facing each other.

28. Which of the following statements is true?

(a) Tantalum capacitors are a type of electrostatic capacitor.
(b) Tantalum capacitors are a type of electrolytic capacitor.

29. Which of the following statements is true regarding aluminum electrolytic capacitors?

(a) The dielectric of the capacitor is aluminum.
(b) The dielectric of the capacitor is an oxide.

30. A mylar capacitor:

(a) is encased in a mylar coating that is sealed with wax.
(b) has a dielectric of mylar film.
(c) is a type of electrolytic capacitor.
(d) is usually not made with a breakdown voltage above 500 volts

31. A 50-μF electrolytic capacitor is purchased, and later it is learned that it actually has a capacitance of 90 μF. Which of the following is true?

(a) The capacitar is out of tolerance.
(b) This could never happen.

(c) It is all right because electrolytic capacitors normally have tolerances of −20 to +100 percent.
(d) The capacitor is incorrectly marked because an electrolytic capacitor cannot have such a high capacitance.

32. Which of the following components is sometimes used to stabilize circuits against temperature changes?

(a) Varistors.
(b) Ferrite beads.
(c) LDR's.
(d) Thermistors.

33. Which of the following is not an advantage of mica capacitors?

(a) High breakdown voltage.
(b) Low loss.
(c) Available in both low and high capacitance values.

34. A certain capacitor has a rating of N150. This means that:

(a) the nominal value of capacitance is 150 pF.
(b) the nominal value of capacitance is 150 microfarads.
(c) the capacitance will decrease 150 parts per million when the temperature increases 1°C.

35. Which of the following types of capacitors is usually polarized?

(a) Electrolytic.
(b) Ceramic.
(c) Glass.
(d) Metallized paper.

36. Which of the following is not an advantage of tantalum over aluminum oxide electrolytics?

(a) Higher capacitance for a given size.
(b) Longer life expectancy.
(c) Longer shelf life.
(d) Much lower cost.

37. Variable capacitors are purposely made nonlinear in order to get a linear dial on the radio. Is this statement true or false?

(a) True
(b) False

38. A measure of how much opposition that a capacitor offers to the flow of dc is its:

(a) load.
(b) ESR.
(c) Vdcw.
(d) specific dielectric capacitance.
(e) Q.

39. The reciprocal of the dissipation factor of a capacitor is the:

(a) Q.
(b) specific dielectric capacitance.
(c) Vdcw.
(d) ESR.

40. To decrease the brightness of the trace on a CRT,

(a) the cathode should be made more negative.
(b) the cathode should be made more positive.

41. It is not necessary to use an ion trap:

(a) with a tube that has electrostatic deflection.
(b) with a tube that has electromagnetic deflection.

(c) with a tube that has electromagnetic focusing.
(d) with aluminized picture tubes.

42. Use of a curved shadow mask in a three-gun color picture tube:

(a) reduces the amount of voltage required for focusing.
(b) reduces the amount of dynamic convergence needed to control the beams.
(c) reduces the amount of grid voltage required to cut off the beam.
(d) has not been done up to this time.

43. Which of the following is not likely to affect color purity?

(a) The position of the deflection yoke.
(b) Stray magnetic fields.
(c) Adjustment of the brightness control.
(d) Adjustment of the purity magnet.

44. When the blue phosphor dot is below the red and green phosphor dots in a three-gun tube, then the blue gun is:

(a) above the red and green guns.
(b) below the red and green guns.

45. Figure 4-18 shows the color code on a disc ceramic capacitor. The capacitance is:

(a) 150 microfarads.
(b) 150 picofarads.
(c) 510 microfarads.
(d) 510 picofarads.

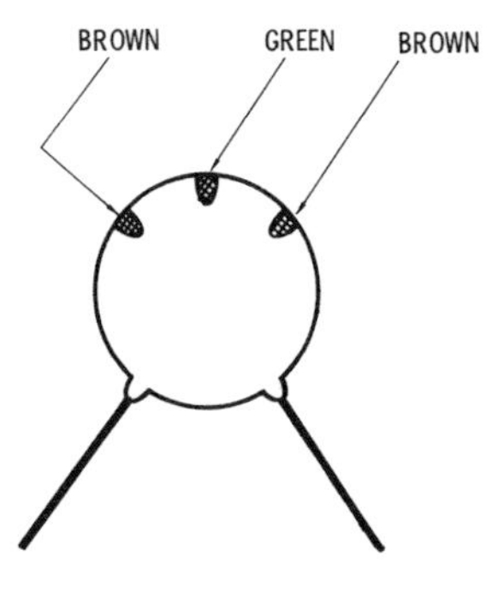

Fig. 4-18. A disc ceramic capacitor with a 3-dot code.

46. For the molded insulated ceramic capacitor of Fig. 4-19, the temperature coefficient is given by the color code on:

(a) A.
(b) C.
(c) D.
(d) E.

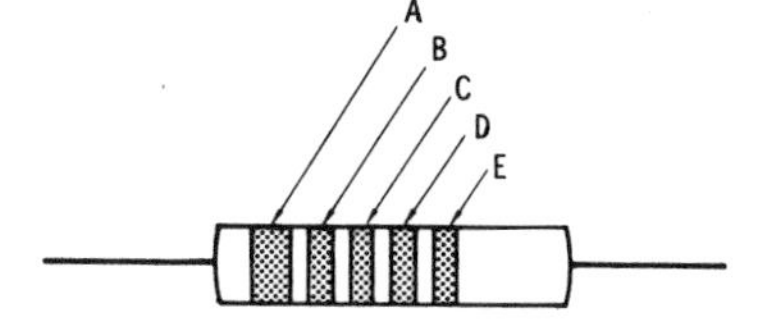

Fig. 4-19. A molded tubular ceramic capacitor with axial leads.

47. An electron beam moves *left and right,* and *up and down* to trace a white rectangle on the face of a CRT. This is called the:

(a) getter.
(b) pincushion.
(c) white square.
(d) raster.

48. In selecting a replacement deflection yoke, a first consideration is the:

(a) size of wire used.
(b) type of core material.
(c) deflection angle.
(d) weight.

49. Two capacitors are placed in parallel. The breakdown-voltage rating of the capacitors in parallel is equal to:

(a) the sum of the breakdown-voltage ratings of each capacitor.
(b) a value that is proportional to the reciprocal of the sums of the breakdown-voltage ratings of each capacitor.
(c) the larger of the two breakdown-voltage ratings.
(d) the smaller of the two breakdown-voltage ratings.

50. A type of microphone that requires a dc current flow through it for its operation is the:

(a) crystal microphone.
(b) carbon microphone.
(c) ribbon microphone.
(d) ceramic microphone.

5

Transistors and Other Semiconductor Devices

KEYED STUDY ASSIGNMENTS

There is no keyed study assignment for this chapter.

IMPORTANT CONCEPTS

In order to pass the CET examination successfully, you must have a working knowledge of transistor and semiconductor theory. This is also true for state and local licensing examinations.

The practice test for this chapter includes questions on general semiconductor theory as well as on the material covered in this chapter. This will give you a chance to test yourself on transistor theory.

Semiconductor Diodes

Breakdown Diodes—Breakdown diodes are designed to be operated with a reverse current. There are two kinds of breakdown diodes in common use. *Zener diodes* conduct current in the reverse direction when they are reverse biased sufficiently to pull electrons out of their covalent bonds. Thus, electron flow in a zener diode is due to the presence of an electric field. *Avalanche diodes* conduct in a manner that is similar to the conduction of current in gas tubes. Collisions of charged particles (either electrons or holes) create additional electron-hole pairs. Fig. 5-1 illustrates the avalanche process.

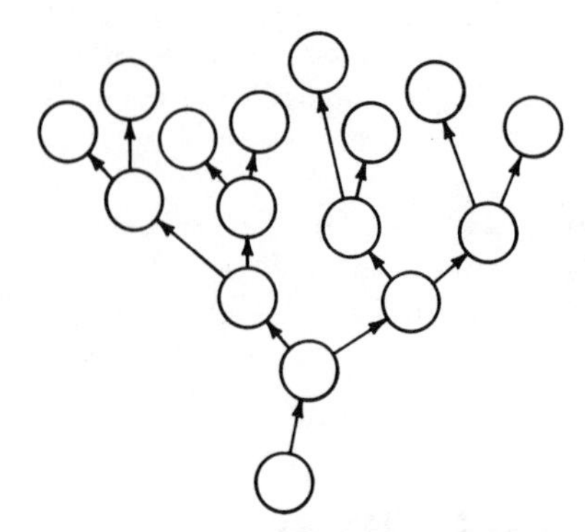

Fig. 5-1. Avalanche current occurs when high-energy charge carriers collide with valance electrons.

Four-Layer Diodes—Four-layer diodes (sometimes called *Shockley diodes*) contain alternate layers of p and n junctions as shown in Fig. 5-2. They act like two transistors which are direct coupled. One transistor is an npn and the other is a pnp.

The four-layer diode is forward biased when a negative voltage is applied to its n-layer terminal and a positive voltage is applied to its p-layer terminal.

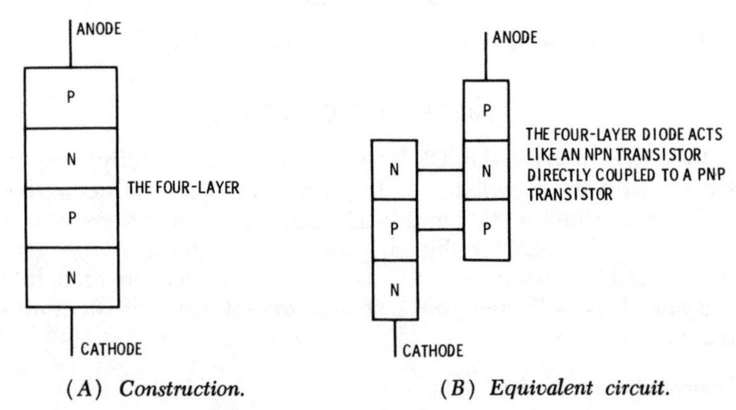

(*A*) *Construction.* (*B*) *Equivalent circuit.*

Fig. 5-2. A four-layer diode.

A small amount of forward bias will start the current flowing. As the forward voltage is increased, a value will be reached which causes the forward current to rise sharply. If the voltage is then reduced somewhat, the high current will still continue to flow. The forward voltage must be dropped to near zero to reduce the current. The action of the four-layer diode is similar to that of a neon lamp, but the voltage required for "firing" is much lower.

Four-layer diodes are useful as switching devices.

Tunnel Diodes—Tunnel diodes (also called *Esaki diodes*) are usually made of heavily doped silicon, although germanium is sometimes used. There is a very thin depletion layer between the p and n junctions. (You will remember that when a p-n junction is formed, there is a movement of charge carriers across the junction, and that this establishes a depletion layer where charge carriers—that is, electrons or holes—do not exist.)

When a very small forward voltage is applied to a tunnel diode, a small forward current flows. As the forward voltage is increased, the forward current drops to zero, and then increases again. This characteristic is shown by the curve of Fig. 5-3.

The region from point *A* to point *B* on the curve of Fig. 5-3 makes it possible for the diode to act as a switch, amplifier, or oscillator.

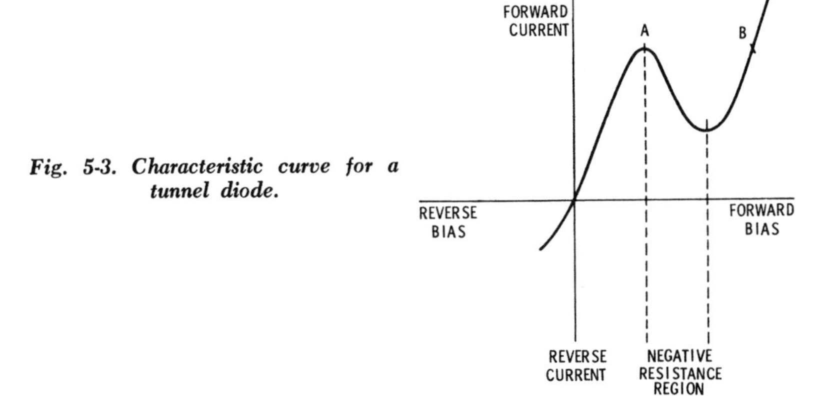

Fig. 5-3. Characteristic curve for a tunnel diode.

Varactor Diodes—Varactor diodes are junction diodes that perform as voltage-variable capacitors. The depletion region of a junction diode increases as the amount of reverse bias is increased. This is shown in Fig. 5-4. The diode acts like a capacitor with the depletion region serving as the dielectric and the charge carriers (holes and electrons) serving as capacitor plates. Moving the plates further apart, by increasing the amount of reverse bias, decreases the capacitance of the device.

Varactors are used as automatic frequency control components, and as part of the resonant-frequency circuit in tuners.

Zener Diodes—Zener diodes are silicon junction diodes that are purposely designed to operate in the reverse-voltage breakdown region of their characteristic curve. (See Fig 5-5.) All junction diodes have a zener voltage. This is the reverse voltage that is suf-

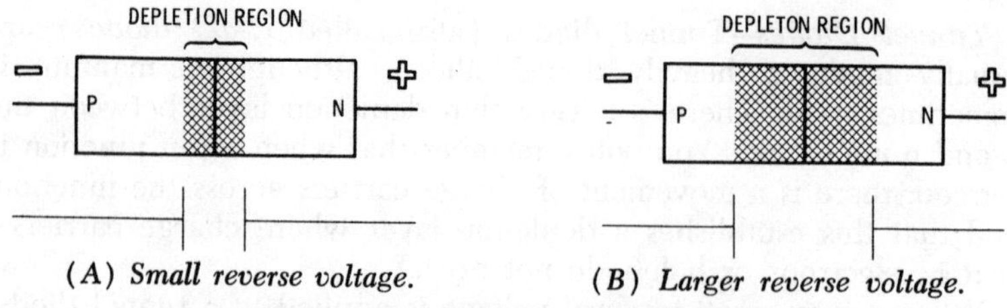

(A) *Small reverse voltage.* (B) *Larger reverse voltage.*

Fig. 5-4. Operation of a varactor diode.

ficiently large to force current through a diode in the reverse direction. While most diodes are destroyed when operated with reverse current, the zener diode is designed to work this way.

As shown in the characteristic curve of Fig. 5-5, when the zener voltage is reached, the voltage across the diode is practically constant regardless of the amount of reverse current flowing. This makes them useful as voltage regulators and as voltage standards. When operated back to back, they can be used for ac regulation.

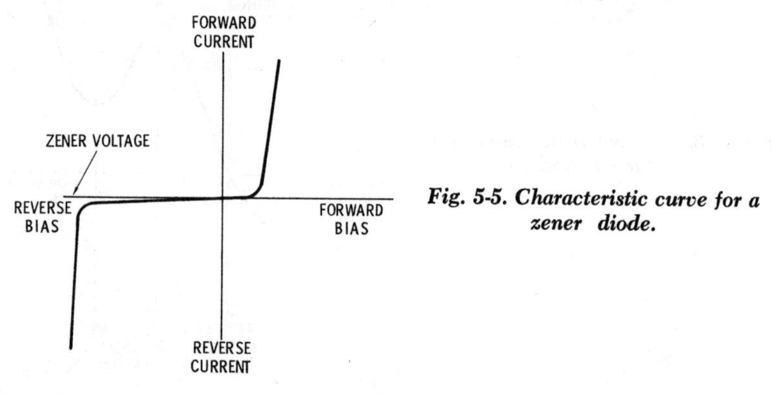

Fig. 5-5. Characteristic curve for a zener diode.

Transistors

Point-Contact Transistors—The first transistors were point-contact types. They were made with two "cat whiskers" touching a piece of semiconductor material.

Point contact transistors had some very important disadvantages: They were not very rugged, they had a low impedance, they had a low maximum operating frequency, they were difficult to make, and they were unstable. Despite these disadvantages, they started the downfall of the vacuum tube.

Grown-Junction Transistors—Grown-junction transistors came next. They were made in two types: npn and pnp. In an npn type, a very thin layer of p-type semiconductor material is sandwiched between two n-type semiconductor layers. A pnp type has a very thin layer of n-type semiconductor material sandwiched between two p-type semiconductor layers. Although they were more rugged, more stable, and easier to make, grown-junction transistors were not much of an improvement as far as frequency limits and impedances are concerned.

Alloy Transistors—Alloy transistors came next. They include the *surface barrier transistor* (*SBT*), *drift transistor, microalloy transistor* (*MAT*), and *microalloy diffused transistor* (*MADT*). The MADT transistor is sometimes called a *surface barrier diffused transistor* (*SBDT*). The order in which these transistors are listed represents a successive improvement in frequency response. Their names come from the manufacturing process, and from the nature of the material used in their fabrication.

Mesa Transistors—Mesa transistors get their name from their appearance. The word "mesa" means table or hill. The mesa transistor is made by selectively etching a silicon or germanium wafer in steps. They appear as a raised plateau over the collector. This type of transistor presents a further improvement in frequency response over previous types.

A further improvement was obtained by adding a very thin resistive layer, called the *epitaxial layer,* between the collector and base. These types of transistors are called *epitaxial mesa transistors.* The epitaxial layer makes it possible to operate the transistor at a higher voltage.

Planar Transistors—Planar transistors have a flat surface (like a plain). They are made by etching and diffusing the emitter and base electrodes on a wafer of p- or n-type material. When the completed transistor is coated with an oxide layer to protect it from contamination, it is said to be *passivated.* An epitaxial layer may be added to increase the collector operating voltage, in which case it is called a *planar epitaxial passivated transistor,* or simply PEP transistor.

Unijunction Transistors—A unijunction transistor has two base electrodes and one emitter. There is no collector. Fig. 5-6 shows that the unijunction transistor has an n-type wafer and a p-type emitter junction. The base leads do not act as rectifying junctions. Instead, they are simply metallized connectors for current flow.

When base number two is positive and base number one is grounded, there is a current through the n-type material. A voltage drop occurs across this material, placing the emitter above ground.

When a small positive voltage is applied to the emitter, the current is not seriously affected. However, as the emitter is made more and more positive, a value will be reached that forces holes into the n region. This lowers the resistance and increases the current through the n-type material.

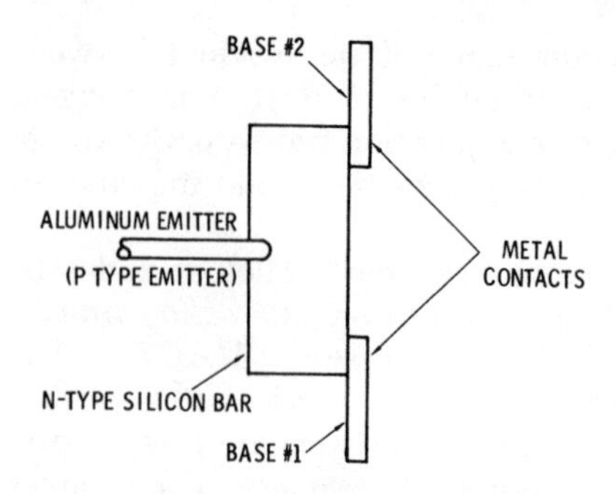

Fig. 5-6. An unijunction diode.

The obvious application for unijunction transistors is as switching devices. A small amount of switching voltage can cause a relatively large amount of current between the bases.

All of the transistors discussed so far are *bipolar*. This simply means that they depend upon the flow of both electrons and holes for their operation. The *field-effect transistor* (*FET*) is a *unipolar* type because it relies upon only one type of charge carrier—either electrons or holes.

Junction Field-Effect Transistors—Fig. 5-7 shows how a junction field-effect transistor (JFET) works. An n-type piece of material is circled at its center by a p-type band. The electrons enter the *source* and arrive at the *drain*. These electrodes are comparable to the cathode and plate of a vacuum tube.

As electrons move from the source to the drain through the *channel,* they encounter an electric field due to the negative voltage on the gate. The more negative this field, the fewer the number of elec-

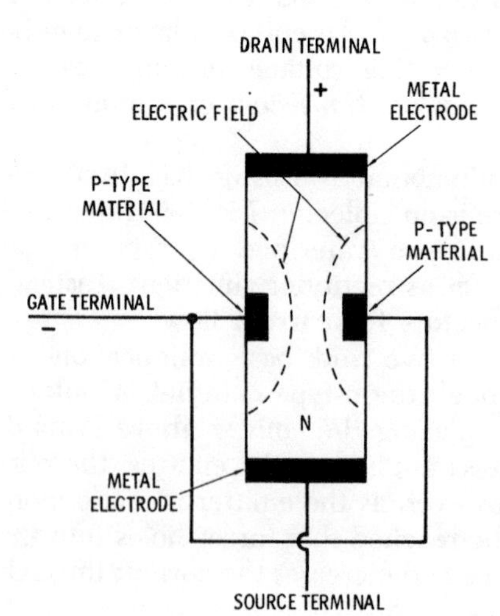

Fig. 5-7. An n-channel JFET.

trons that can pass the gate. This is similar to using a negative voltage on the grid of a tube to control the number of electrons that can flow from cathode to plate. If the negative voltage is sufficiently large, no current will flow from the source to the drain. The voltage required for cutting off the current is called the *pinch-off voltage.*

Instead of using an n-type channel, the channel can be made of p-type material. Of course, the voltages applied to a p-channel FET are opposite in polarity to those for an n-channel FET.

The JFET conducts current when there is no gate voltage. In the absence of a gate voltage, the JFET acts like a resistor. As the gate voltage is added (of proper polarity) the source-to-drain current is reduced (that is, depleted). For this reason, these FET's are called *depletion type* or *depletion mode* field-effect transistors.

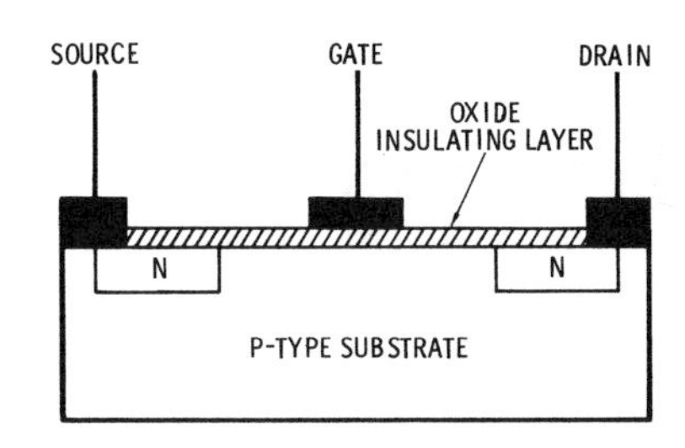

Fig. 5-8. A MOSFET.

A small amount of forward bias on the gate of a JFET will cause an exponential rise of gate current. That is why the gate junction is operated with reverse bias. A small amount of leakage current exists with reverse bias.

The leakage-current characteristic of JFET's can be eliminated by adding an insulating layer between the gate and the channel. Fig. 5-8 shows this type of FET. Because of the insulating layer, it is called an *insulated gate FET (IGFET)*. Since an oxide is used for the insulating layer, this type of transistor is also called *metal-oxide semiconductor FET (MOSFET)*. As with the JFET, a MOSFET can be either an n-channel or p-channel type.

There are two ways to make a MOSFET. If zero volts on the gate does not prevent current from flowing through the channel, it is a depletion type. This type of operation is the same as for the JFET. If there is no current through the MOSFET when the gate voltage is zero, and adding a gate voltage (of proper polarity) causes current to flow, it is called an *enhancement type* (or, *enhancement mode*) MOSFET. The two types of MOSFET's are illustrated in Fig. 5-9.

The MOSFET has the obvious advantage that there is isolation between the gate and channel. It is the field from the gate that controls the current through the channel. The insulation layer reduces the gate current to zero (for all practical purposes), and therefore, the MOSFET can be operated with forward voltage on the gate. A disadvantage of the MOSFET is its characteristic high noise. Another disadvantage is that the MOSFET can be destroyed by a static charge. As an extra precaution, they should be stored with their leads shorted together.

More than one gate can be used for controlling the flow of current in the FET. Fig. 5-10 shows a *dual-gate MOSFET*.

Thyristors

A thyratron is a gas-filled triode tube. When the plate of a thyratron is positive with respect to its cathode, it can be triggered into conduction. A positive pulse on the grid "fires" the thyratron into

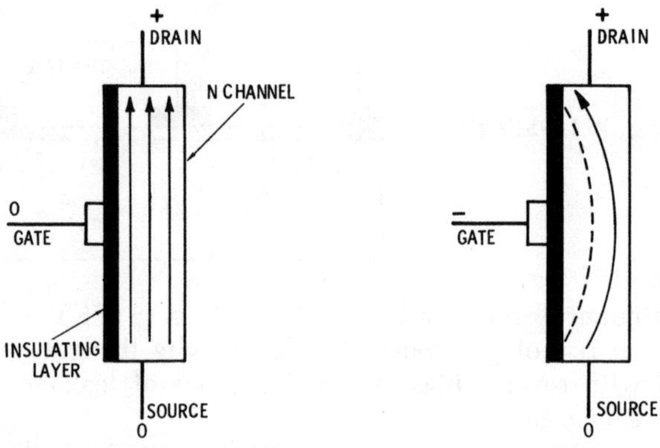

(A) Depletion type—high conduction. *(B) Depletion type—low conduction.*

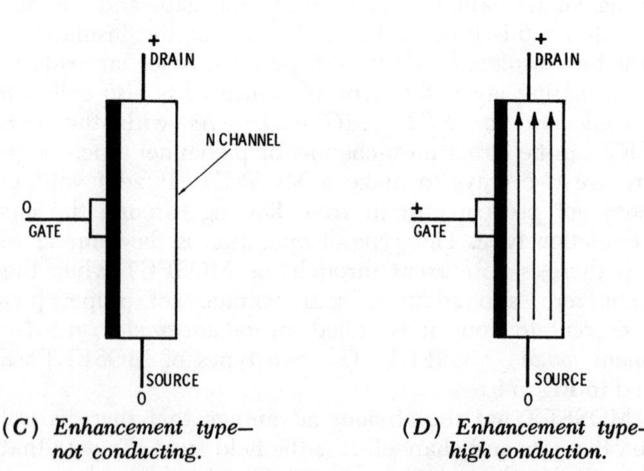

(C) Enhancement type—not conducting. *(D) Enhancement type—high conduction.*

Fig. 5-9. Two types of n-channel FET's.

conduction, and once the tube conducts, it cannot be cut off with a negative grid voltage.

Thyristors are semiconductor devices that have characteristics similar to those of a thyratron. Examples are the silicon-controlled rectifier (SCR), the triac, and the diac.

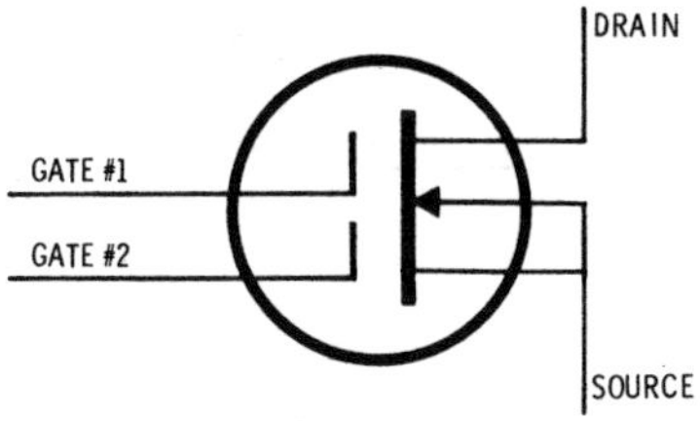

Fig. 5-10. A dual-gate MOSFET.

DRAIN
GATE #1
GATE #2
SOURCE

Silicon Controlled Rectifiers (SCR)—Fig. 5-11 shows the symbol for an SCR. As with any diode, it is forward biased when its anode is positive with respect to its cathode. With a small amount of forward bias the SCR will conduct. As the forward bias is increased, a point will be reached where conduction increases rapidly. This is called the *breakover point.* A positive current pulse on the gate of the SCR reduces the breakover voltage. In other words, a positive current pulse will turn the SCR on. Once the SCR conducts, it cannot be shut off by the gate.

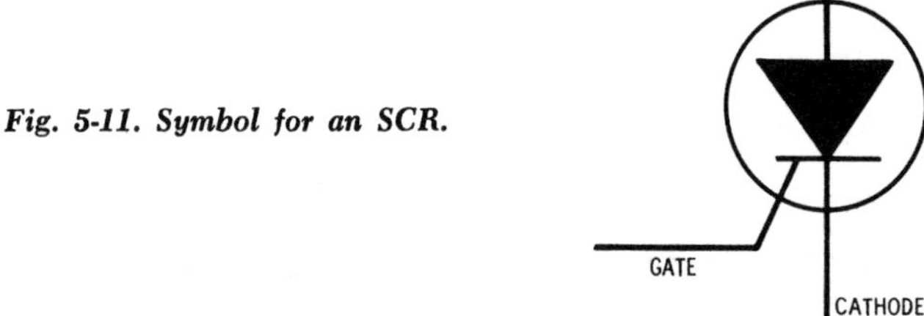

Fig. 5-11. Symbol for an SCR.

Triacs—A triac is the equivalent of two SCR's connected as shown in Fig. 5-12. The triac is different from an SCR by the fact that it can conduct in either direction, and also by the fact that it can be turned on by either a positive or a negative current pulse.

Diacs—A diac is a three-layer diode. Fig. 5-13 shows how it is made. Note that it is similar to a bi-polar transistor except that there is no connection made to the section that would be the base. Diacs can conduct in either direction, provided the voltage across the device is above the breakover point. Once the breakover point is

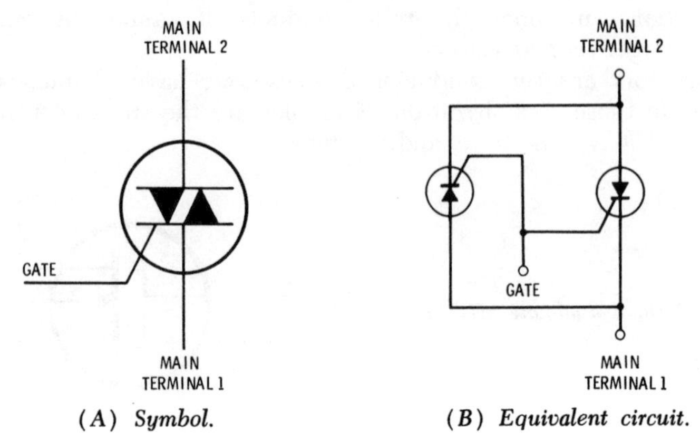

(A) *Symbol.* (B) *Equivalent circuit.*

Fig. 5-12. A triac.

reached, avalanche breakdown occurs and the diac conducts. It will continue to conduct even though the forward bias is reduced somewhat below the breakover voltage value.

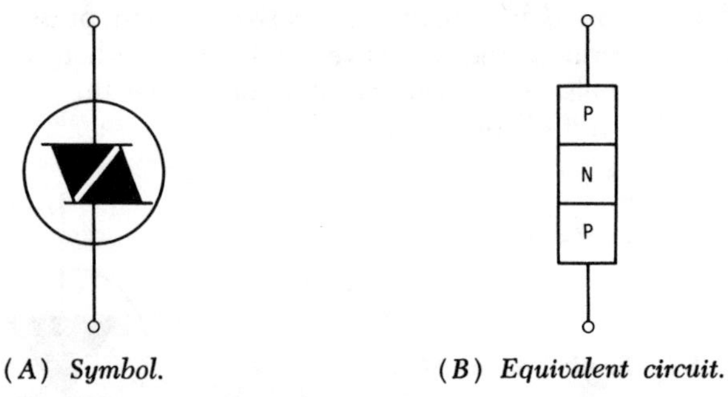

(A) *Symbol.* (B) *Equivalent circuit.*

Fig. 5-13. The diac.

PROGRAMMED QUESTIONS AND ANSWERS

Starting with question number 1, select the answer that you feel is correct. If you feel that (A) is correct, proceed to block number 17 as directed. If you feel that (B) is correct, proceed to block number 9 as directed. If you feel that more than one answer is correct, choose the one that you think is the *most* correct.

. . . .

1 Which of the following types of diodes is known for its ability to switch very rapidly, amplify, or operate as an oscillator?

(A) Varactor diode. (Go to block number 17.)

(B) Tunnel diode. (Go to block number 9.)

2 Your answer is wrong. The agc circuit does not require the use of a voltage-variable capacitor. Go to block number 14.

3 Alloy types offer a definite improvement in frequency response over grown-junction types.

Here is your next question . . .

Which of the following types of transistors can be operated at a higher voltage?

(A) Surface barrier. (Go to block number 7.)

(B) Epitaxial mesa. (Go to block number 19.)

4 Your answer is wrong. Esaki diode is another name for a tunnel diode. Go to block number 12.

5 Your answer is wrong. The upper frequency limit on the operation of early point-contact transistors was one of their biggest disadvantages. Go to block number 25.

6 Your answer is wrong. The symbol shown in Fig. 5-15B is for a p-channel depletion type MOSFET. Note the gate symbol which indicates that it is insulated from the channel. Remember that the MOSFET is also known as an insulated-gate FET. Go to block number 18.

7 Your answer is wrong. Go to block number 19.

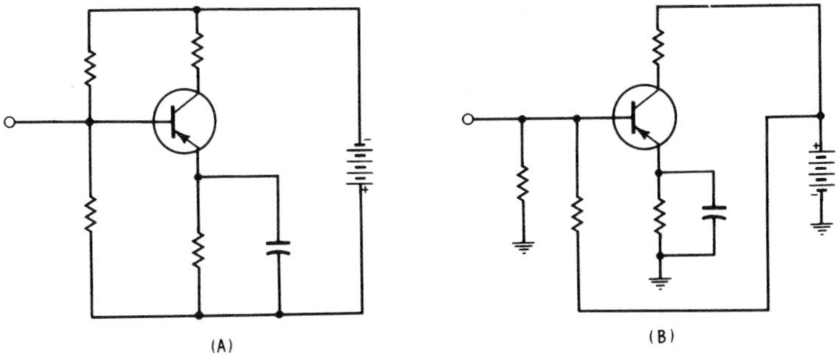

(A) (B)

Fig. 5-14. Which of the amplifiers is incorrectly biased?

. . .

8 The dual-gate FET is used in the tuner as an rf amplifier. One gate is used for agc control, and the other gate has the rf signal applied to it.

Here is your next question . . .

Which of the amplifiers in Fig. 5-14 is improperly biased?

(A) The amplifier shown in (A). (Go to block number 13.)

(B) The amplifier shown in (B). (Go to block number 22.)

. . . .

9 The tunnel diode is used as a switch, as an amplifier, and as an oscillator.

Here is your next question . . .

Which of these circuits would be more likely to use a varactor?

(A) Afc. (Go to block number 14.)

(B) Agc. (Go to block number 2.)

. . . .

10 Your answer is wrong. Unijunction transistors are used in switching applications. Go to block number 8.

. . . .

11 The pinchoff voltage for a p-channel JFET is positive, which is opposite to the polarity required for pinchoff in an n-channel JFET.

Here is your next question . . .

Which of the following will conduct equally well in two directions?

(A) Triacs. (Go to block number 16.)

(B) Bipolar transistors. (Go to block number 24.)

. . . .

12 Zener diodes are a type of breakdown diode. They are used for establishing a voltage reference because the voltage across a zener diode is relatively independent of current when it is operated within the manufacturer's specifications.

Here is your next question . . .

Which of the following types of diodes acts like two direct-coupled transistors?

(A) An avalanche diode. (Go to block number 23.)

(B) A four-layer diode. (Go to block number 15.)

. . . .

13 Your answer is wrong. A pnp transistor requires a negative voltage on its collector (with respect to the voltage on its emitter). Go to block number 22.

. . . .

14 In the afc (automatic frequency control) circuit a dc correction voltage is used to maintain the local oscillator on frequency. This correction voltage determines the capacitance of a varactor in the oscillator tuned circuit.

Here is your next question . . .

Which of the following is classified as a "breakdown diode"?

(A) A zener diode. (Go to block number 12.)

(B) An Esaki diode. (Go to block number 4.)

. . . .

15 The four-layer diode contains alternate layers of n- and p-type material. It acts like two direct-coupled transistors. In circuits its action is analogous to that of a neon lamp, except that it "fires" at a much lower voltage.

Here is your next question . . .

Which of the following was an important advantage of early point-contact transistors?

(A) They were able to operate at very high frequencies. (Go to block number 5.)

(B) They did not require a filament voltage. (Go to block number 25.)

. . . .

16 A triac is like an SCR except that it can conduct in either of two directions.

Here is your next question . . .

In an enhancement mode n-channel MOSFET, should a positive or negative voltage be applied to the gate to obtain conduction? (Go to block number 26.)

. . . .

17 Your answer is wrong. A varactor is a voltage-variable capacitor. It is used in tuned circuits. Go to block number 9.

. . . .

18 The arrow points in for an n-channel JFET symbol.

Here is your next question . . .

For a p-channel JFET the pinchoff voltage on the gate would be:

(A) negative with respect to the voltage on the source. (Go to block number 20.)

(B) positive with respect to the voltage on the source. (Go to block number 11.)

19 The epitaxial layer makes it possible to operate the transistor at a higher voltage.

Here is your next question . . .

Which of the following types of transistors is more likely to be used as an rf amplifier in a color receiver?

(A) Unijunction. (Go to block number 10.)

(B) Dual-gate FET. (Go to block number 8.)

20 Your answer is wrong. Remember this important fact: The voltages on a depletion-type n-channel FET are like those on a triode tube. Thus, a sufficient amount of negative voltage will cut off a triode and an n-channel depletion-type FET. However, the voltages on a p-channel depletion-type FET would be opposite to those on a triode Go to block number 11.

21 Your answer is wrong. The alloy types offer a definite improvement in frequency response over the grown-junction type. Go to block number 3.

22 The pnp transistor will not operate properly with a positive voltage (with respect to the emitter) on its collector.

Here is your next question . . .

Which of the symbols in Fig. 5-15 is the symbol for an n-channel junction-type FET?

(A) The symbol shown in Fig. 5-15A is for an n-channel JFET. (Go to block number 18.)

(B) The symbol shown in Fig. 5-15B is for an n-channel JFET. (Go to block number 6.)

23 Your answer is wrong. An avalanche diode is a form of breakdown diode. Go to block number 15.

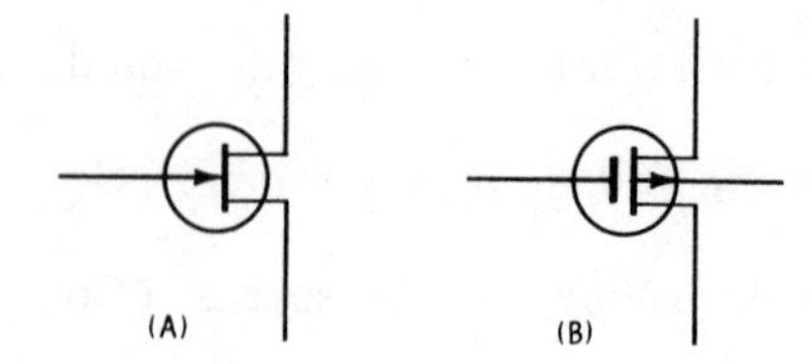

Fig. 5-15. Which is the correct symbol for an n-channel JFET?

. . . .

24 Your answer is wrong. A bipolar transistor conducts by holes *and* electrons, while a unipolar transistor, such as an FET) conducts by holes *or* electrons. Go to block number 16.

. . . .

25 The fact that transistors do not require a filament power supply has always been an advantage.

Here is your next question . . .

Which of the following types of transistors would normally have the higher frequency response?

(A) Alloy type. (Go to block number 3.)

(B) Grown junction type. (Go to block number 21.)

. . . .

26 A positive voltage is required.

You have now completed the programmed questions and answers.

PRACTICE TEST

1. Which of the following types of diodes is sometimes used as an amplifier?

(a) Zener diodes.
(b) Tunnel diodes.
(c) Junction diodes.
(d) Diacs.

2. Which of the following is most nearly like a thyratron in its operation?

(a) MOSFET.
(b) MADT.
(c) SCR.
(d) Diac.

3. Which of the following is a type of breakdown diode?

(a) Zener diode.
(b) Tunnel diode.
(c) Junction diode.
(d) LDR.

4. Which of the following is a bipolar transistor?

(a) Diac.
(b) MOSFET.
(c) JFET.
(d) Npn.

5. Which of the following is most likely to be destroyed by an accidental static charge from the probe of a meter?

(a) MOSFET.
(b) JFET.
(c) SCR.
(d) Npn.

6. The electrode on a JFET that corresponds to the grid of a triode is the:

(a) source.
(b) gate.
(c) drain.
(d) base.

7. The electrode on a JFET that corresponds to the plate of a triode is the:

(a) source.
(b) gate.
(c) drain.
(d) collector.

8. The electrode on a JFET that corresponds to the cathode of a triode is the:

(a) source.
(b) gate.
(c) drain.
(d) emitter.

9. Which of the following can conduct in either of two directions?

(a) Bipolar transistors.
(b) SCR's.
(c) Triacs.
(d) Npn transistors.

10. Which of the following would be most suitable for use as an amplifier?

(a) Triac.
(b) Diac.
(c) SCR.
(d) IGFET.

11. Which electrode of an SCR corresponds to the grid of its equivalent tube device?

(a) Anode.
(b) Drain.
(c) Gate.
(d) Emitter.

12. Another name for tunnel diode is:

(a) Esaki diode.
(b) Shockley diode.
(c) diac.
(d) breakdown diode.

13. The region on both sides of a p-n junction in which charge carriers do not exist is called the:

(a) no-man's land.
(b) Barren Straits.
(c) waste region.
(d) depletion region.

14. Which of the following diodes exhibits a negative resistance characteristic as its forward bias is increased from zero?

(a) Esaki diode.
(b) Shockley diode.
(c) Diac.
(d) Breakdown diode.

15. Increasing the reverse bias on a varactor diode will:

(a) decrease its capacitance.
(b) increase its capacitance.
(c) not affect its capacitance.

16. A reverse bias on a p-n junction diode will normally destroy the diode if:

(a) the reverse bias is pulsed.
(b) the reverse bias reaches the zener value.
(c) the reverse bias is steady over a long period of time.
(d) the reverse bias is pure dc.

17. A four-layer diode is sometimes called:

(a) an Esaki diode.
(b) a Shockley diode.
(c) a diac.
(d) a passivated diode.

18. The operation of a four-layer diode is most nearly like the action of:

(a) a phanatron.
(b) a thyratron.
(c) an ignitron.
(d) a neon lamp.

19. Which of the following is a form of breakdown diode?

(a) Tunnel diode.
(b) Avalanche diode.
(c) P-n junction diode.
(d) LDR.

20. Which of the following is not a disadvantage of point-contact transistors when compared to drift transistors?

(a) Their upper-frequency limit of operation is not as great.
(b) They are not as rugged.
(c) They require a much larger power for their operation.

21. Which of the following is not a type of alloy transistor?

(a) SBT.
(b) MAT.
(c) MADT.
(d) PEP.

22. Mesa transistors get their name from:

(a) the name of the man who invented them.
(b) the name of the company that makes them.
(c) the type of junction used.
(d) their shape.

23. Higher-voltage operation of a transistor is obtained by the use of:

(a) dopants.
(b) a passivated layer.
(c) a substrate.
(d) an epitaxial layer.

24. A planar transistor is protected from contamination by the use of:

(a) dopants.
(b) a passivated layer.
(c) a substrate.
(d) an epitaxial layer.

25. Which of the following types of transistors does not have a collector?

(a) Unijunction transistor.
(b) PEP transistor.
(c) MADT.
(d) Junction transistor.

26. Which of the following types of transistors is not bipolar?

(a) MADT.
(b) PEP transistor.
(c) SBT.
(d) IGFET.

27. Which of the following types of transistors may be operated in the enhancement mode?

(a) MADT.
(b) PEP.
(c) Drift transistor.
(d) IGFET.

28. The pinchoff voltage on the gate of a p-channel FET will be:

(a) negative with respect to the voltage on the source.
(b) positive with respect to the voltage on the source.

29. The voltage on the collector of an npn transistor used as an audio amplifier is positive, and the voltage on the base is negative with respect to the emitter. Which of the following is true?

(a) This is normal.
(b) The polarity of the voltage on the collector is wrong.
(c) The polarity of the voltage on the base is wrong.
(d) The polarities of both voltages are wrong.

30. Which of the following is the same as an IGFET?

(a) JFET.
(b) MOSFET.

31. Static charges are most likely to destroy:

(a) a PEP transistor.
(b) an SBT.
(c) a MADT.
(d) a MOSFET.

32. Which of the following is not a thyristor?

(a) A triac.
(b) A diac.
(c) An Esaki diode.
(d) An SCR.

33. A small amount of positive current in the gate of an SCR will:

(a) increase the breakover voltage.
(b) decrease the breakover voltage.
(c) decrease the leakage current.
(d) prolong the life of the SCR.

34. A three-layer diode is known as:

(a) a Shockley diode.
(b) a tunnel diode.
(c) a diac.
(d) an LDR.

35. Your measurements show that current is flowing from the anode to the cathode in a zener diode. This is:

(a) an indication that the diode is in the circuit backwards.
(b) the proper direction of current flow.

36. A tunnel diode may be used as:

(a) a voltage regulator.
(b) a high-voltage rectifier.
(c) a detector.
(d) an oscillator.

37. The majority charge carriers in p-type material are:

(a) holes.
(b) electrons.
(c) neutrons.
(d) positrons.

38. To reverse bias a semiconductor diode:

(a) its anode is made positive with respect to its cathode.
(b) its anode is made negative with respect to its cathode.

39. Which of the following types of semiconductor devices would be used for obtaining a constant value of voltage between two points?

(a) A diac.
(b) A tunnel diode.
(c) A zener diode.
(d) A four-layer diode.

40. Which of the following types of diodes would be most useful in a ratio-detector circuit?

(a) Four-layer diodes.
(b) Three-layer diodes.
(c) Varactor diodes.
(d) P-n junction diodes.

41. Which of the following is NOT true regarding the circuit of Fig. 5-16?

(a) It is a grounded-base circuit.
(b) It has less than unity current gain and low input resistance.
(c) It does not produce a 180° phase inversion between the input and output signals.
(d) It will not work because the voltage polarity is wrong.

42. **Which of the following devices operates like two direct-coupled transistors?**

(a) A tunnel diode.
(b) A four-layer diode.
(c) A zener diode.
(d) A varactor diode.

43. **Which of the symbols shown in Fig. 5-17 is for a p-channel JFET?**

(a) The symbol marked (a).
(b) The symbol marked (b).
(c) The symbol marked (c).

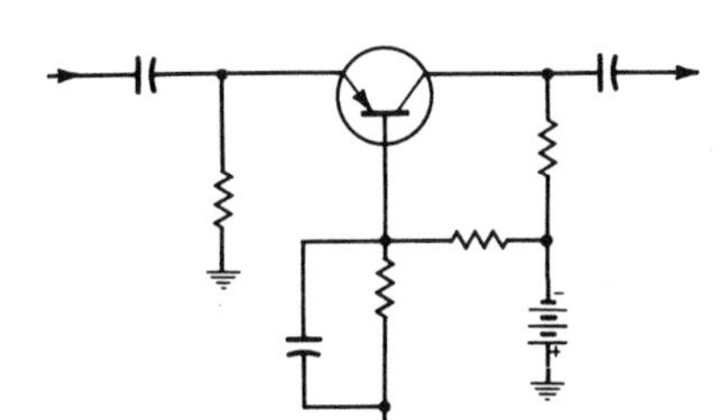

Fig. 5-16. A transistor amplifier.

44. **The base-collector junction of a transistor is:**

(a) always passivated.
(b) forward biased.
(c) reverse biased.
(d) never a grown junction.

45. **The current gain in a common-emitter transistor configuration is called:**

(a) the alpha.
(b) the beta.
(c) the gain-bandwidth product.
(d) the impedance transfer.

46. **The current gain in a common-base transistor configuration is called:**

(a) the alpha.
(b) the beta.
(c) the gain-bandwidth product.
(d) the impedance transfer.

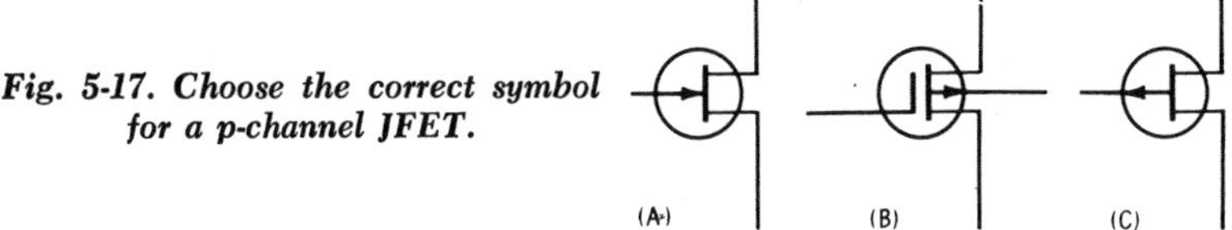

Fig. 5-17. Choose the correct symbol for a p-channel JFET.

47. **The frequency at which the common-base current gain drops to 70.7 percent of its value at 1 kHz is called the:**

(a) alpha cutoff frequency.
(b) beta cutoff frequency.
(c) gain-bandwidth product.
(d) admittance transfer frequency.

48. **The frequency at which the common-emitter forward-current transfer ratio is unity is the:**

(a) alpha cutoff frequency.
(b) beta cutoff frequency.
(c) gain-bandwidth product.
(d) admittance transfer frequency.

49. The frequency at which the common-emitter current gain drops to 70.7 percent of its value at 1 kHz is called the:

(a) alpha cutoff frequency.
(b) beta cutoff frequency.
(c) gain-bandwidth product.
(d) admittance transfer frequency.

50. With respect to the voltage on the drain, the polarity of voltage on the gate of an n-channel JFET is normally

(a) positive.
(b) negative.

6

Basic Mathematics and Circuit Analysis

KEYED STUDY ASSIGNMENT

Howard W. Sams Photofact Television Course
Chapter 5—Resistance-Capacitance Circuit Characteristics

As a service technician, you have no doubt encountered Ohm's law in your study of electronics. Occasionally, you may combine resistors or capacitors in order to get a desired value. You should have an understanding of impedance, reactance, Q, etc. In this chapter some of the basic mathematics relating to electronics will be reviewed.

A good knowledge of the mathematics gives insight to the way circuits work. For example, knowing the equation for capacitive reactance:

$$\left(X_c = \frac{1}{2\pi fC}\right)$$

explains why the reactance decreases when the frequency increases. (When you increase the denominator of a fraction, you decrease the value of the fraction.)

In order to pass the CET test or licensing examinations you should know the equations and be able to work problems in all of the areas discussed in this chapter.

IMPORTANT CONCEPTS

Capacitors and Capacitive Reactance

The capacitive reactance in ohms is given by the equation:

$$X_c = \frac{1}{2\pi fC}$$

This equation shows that increasing the frequency or the capacitance will reduce the capacitive reactance.

Capacitors in parallel add their capacitance values, but in series they combine like resistors in parallel. However, *capacitive reactances* in series add, and in parallel they combine like resistances and inductive reactances in parallel.

An important thing to remember about capacitors is that moving the plates closer together increases the capacitance and reduces the reactance. Increasing the area of the plates, or using a material with a higher dielectric constant, also increases the capacitance and reduces the reactance.

When capacitors are placed in series across a voltage, the largest voltage drop appears across the smallest capacitor. This is easy to remember because the smallest capacitance will produce the largest capacitive reactance.

Impedance

If you are unsure of yourself on square roots, you can solve impedance problems graphically. It is necessary only to remember the impedance triangle of Fig. 6-1. If inductive reactance (X_L) and capacitive reactance (X_c) are both given, subtract the smaller from the larger to get the net reactance (X). Draw a line that has a length proportional to R, and at the end draw another line at a right angle (90°) that is proportional to X. The impedance line (Z) will have a length that is proportional to the impedance in ohms.

Many impedance problems are based on standard triangles. These are triangles that have standard ratios and angles. Fig. 6-2 shows three of the most commonly used standard triangles. Knowing these triangles can simplify the solution of impedance problems considerably.

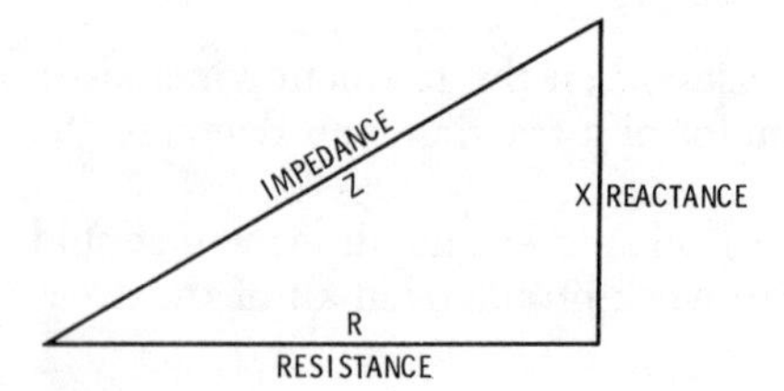

Fig. 6-1. The impedance triangle.

Sample Problem—A coil with an inductive reactance of 300 ohms is in series with a 400-ohm resistor. What is the impedance of the combination?

Solution—Mathematically this problem can be solved as follows:

$$\begin{aligned} Z &= \sqrt{R^2 + (X_L - X_c)^2} \\ &= \sqrt{(400)^2 + (300)^2} \quad \text{(Note that } X_c = 0.\text{)} \\ &= \sqrt{160{,}000 + 90{,}000} \\ &= \sqrt{250{,}000} \\ &= 500 \text{ ohms} \quad \text{(Answer)} \end{aligned}$$

If you note that the reactance and resistance are in a ratio of 3 to 4, you immediately know that a 3-4-5 triangle (Fig. 6-2) is involved. You can immediately write the answer—*500 ohms.*

Sample Problem—A capacitor with a reactance of 100 ohms is in series with a 100-ohm resistor. What is the circuit impedance?

Solution—Mathematically this problem can be solved as follows:

$$\begin{aligned} Z &= \sqrt{R^2 + (X_L - X_c)^2} \\ Z &= \sqrt{100^2 + 100^2} \quad \text{(Note that } X_L = 0.\text{)} \\ &= \sqrt{10{,}000 + 10{,}000} \\ &= \sqrt{20{,}000} \\ &= 141.4 \text{ ohms} \quad \text{(Answer)} \end{aligned}$$

Again, if you know the standard triangles, this problem can be solved by inspection. Note that the resistance and reactance are in a

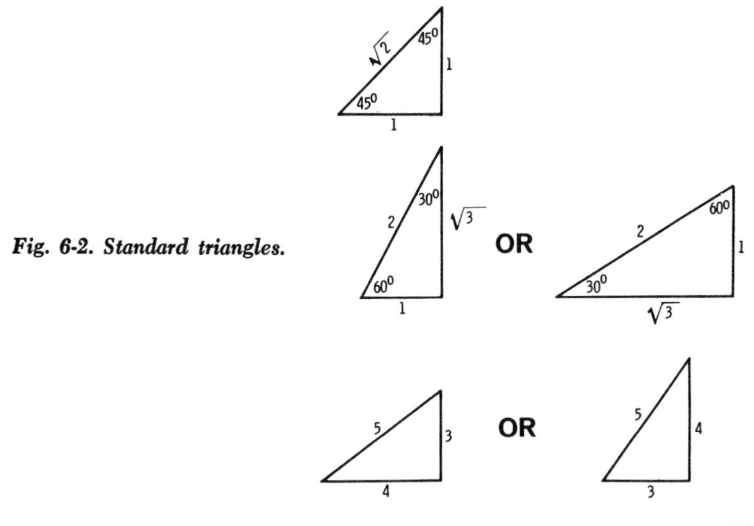

Fig. 6-2. Standard triangles.

ratio of one to one. The 45° standard triangle shows that the impedance will be 1.414 times the resistance (or reactance) under this condition. Thus, the value of the impedance, 141.4 ohms, can be written by inspection.

A final word on impedance calculations: *Don't* mix units. If the resistance is in kilohms and the reactance is in ohms, convert them both to ohms before solving the problem.

Inductors and Inductive Reactance

The inductive reactance in ohms is given by the equation:

$$X_L = 2\pi fL.$$

This equation shows that increasing the frequency or the inductance will increase the inductive reactance.

Inductances and inductive reactances in series add provided there is no mutual inductance between the coils. Inductances and inductive reactances in parallel combine in the same way as resistances in parallel. Again, this assumes that there is no mutual inductance between the coils.

Moving the turns of a coil closer together, or using a core material with a higher permeability will increase the inductance of a coil and increase its inductive reactance.

Kirchhoff's Laws

Kirchhoff's current law states that the sum of the currents entering a junction equals the sum of the currents leaving that junction. Kirchhoff's voltage law states that the algebraic sum of the voltage drops in a circuit equals the sum of the voltage rises. Knowledge of these laws may reduce the number of measurements needed when troubleshooting.

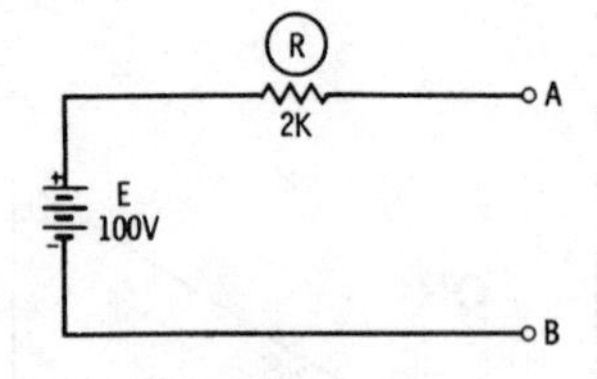

Fig. 6-3. Determine the voltages between A and B.

Sample Problem—What is the voltage between *A* and *B* in the circuit of Fig. 6-3?

Solution—With no current, the voltage drop across R is zero. Therefore, there must be 100 volts between *A* and *B*.

Sample Problem—In the circuit of Fig. 6-4 find the base current.

Solution—The emitter current is found by Ohm's law:

$$\text{Emitter Current} = \frac{\text{Emitter Voltage}}{\text{Emitter Resistance}}$$

$$= \frac{1}{10^3}$$

$$= 1 \text{ mA}$$

There is 1 mA entering the emitter, and 0.95 mA leaving the collector. According to Kirchhoff's current law, the total current leaving the transistor must equal the total current entering.

$$I_E = I_B + I_C$$

$$1 \text{ mA} = I_B + 0.95 \text{ mA}$$

$$I_B = 1 \text{ mA} - 0.95 \text{ mA}$$

$$= 0.05 \text{ mA} \qquad \text{(Answer)}$$

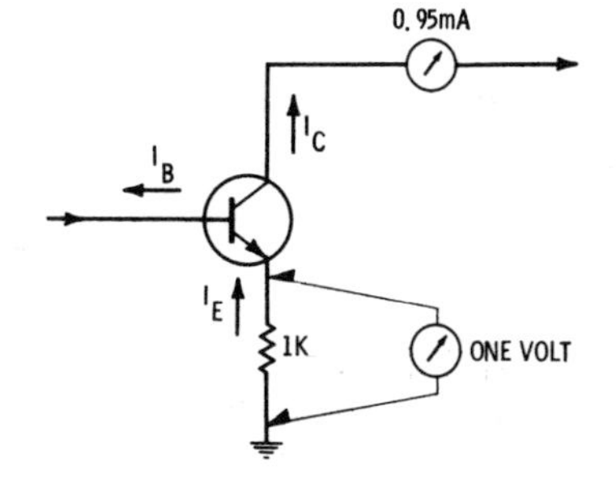

Fig. 6-4. Use Kirchhoff's current law to determine base current.

Ohm's Law

For the purpose of taking CET or licensing examinations, you are expected to be thoroughly familiar with Ohm's law. It is sometimes stated mathematically in two different forms.
For dc circuits:

$$I = \frac{E}{R}.$$

For ac circuits:

$$I = \frac{E}{Z},$$

where,

$$Z = \sqrt{R^2 + (X_L - X_C)^2}$$

For ac circuits E and I are normally rms (effective) values.

You must be careful not to mix units when solving an Ohm's law problem. There are a lot of rules relating to units. For example:

$$\text{Current in Milliamperes} = \frac{\text{Voltage in Volts}}{\text{Resistance in Kilohms}}$$

However, you will usually be better off to convert all of the measurements to the basic units of volts, amperes, and ohms. Then you can convert your answer to whichever unit is convenient.

Sample Problem—What is the voltage drop across a two-megohm resistor when a current of one milliampere is following through it?

Solution—Two megohms can be written as 2×10^6 ohms, or as 2,000,000 ohms. A current of one milliampere can be written as 1×10^{-3} ampere or as 0.001 ampere. By Ohm's law:

$$\begin{aligned} E &= IR \\ &= 2 \times 10^6 \times 1 \times 10^{-3} \\ &= 2 \times 10^3 \\ &= 2000 \text{ volts} \qquad \text{(Answer)} \end{aligned}$$

or, without powers of 10:

$$\begin{aligned} E &= IR \\ &= 2{,}000{,}000 \times 0.001 \\ &= 2000 \text{ volts} \qquad \text{(Answer)} \end{aligned}$$

You may not have to work an Ohm's law problem, but may have a problem that requires the use of Ohm's law to get the answer.

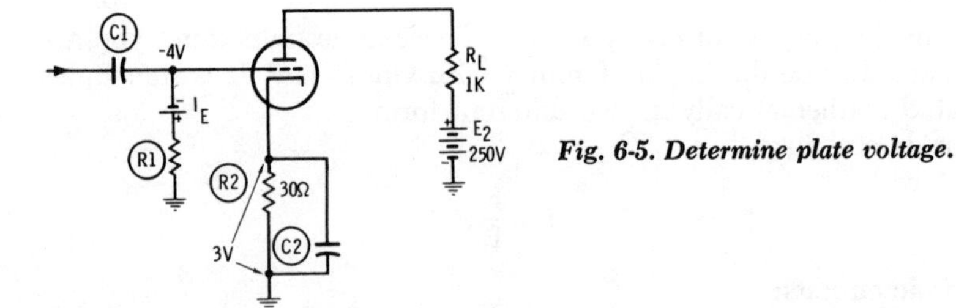

Fig. 6-5. Determine plate voltage.

Sample Problem—In the circuit of Fig. 6-5, what is the voltage on the tube plate with respect to the chassis?

Solution—The voltage on the plate will be equal to the plate supply voltage (E_2) minus the drop across the plate load (R_L). The first step is to find the plate current which, in this case, is also the cathode current.

$$I_K = \frac{E_K}{R}$$
$$= \frac{3}{30}$$
$$= 0.1 \text{ ampere}$$

The drop across R_L can be found by multiplying the cathode current (i.e., the plate current) by the plate load resistance.

$$E_L = I_K R_L$$
$$= 0.1 \times 1000$$
$$= 100 \text{ volts}$$

The voltage on the plate is the applied voltage minus the voltage drop across R_L:

$$E_P = E_2 - E_L$$
$$= 250 - 100$$
$$= 150 \text{ volts} \qquad \text{(Answer)}$$

Peak, RMS, and Average Values

For a pure sinusoidal voltage or current, the following equations apply:

$$\text{Average Current} = 0.636 \times \text{Peak Current}$$
$$\text{Average Voltage} = 0.636 \times \text{Peak Voltage}$$
$$\text{RMS Current} = 0.707 \times \text{Peak Current}$$
$$\text{RMS Voltage} = 0.707 \times \text{Peak Voltage}$$

Meters usually give the rms value of voltage or current. However, the meter readings are correct only if a sine-wave voltage or current is being measured.

Power and Power Factor

The power dissipated by current flowing through a resistance is given by the equations:

$$P = I^2R$$
$$P = \frac{E^2}{R}$$
$$P = EI$$

With alternating current, the rms values of voltage and current are used for finding power.

When there is inductance or capacitance in an ac circuit, the power factor is an important consideration. The power factor is the cosine of the phase angle between the voltage and current in a circuit.

The *apparent power* dissipated in an ac circuit is found by multiplying the rms voltage by the rms current. The actual power (called the *true power*) will be less than the apparent power. The power triangle of Fig. 6-6 shows the relationship between the true power, apparent power, and reactive voltamperes (vars).

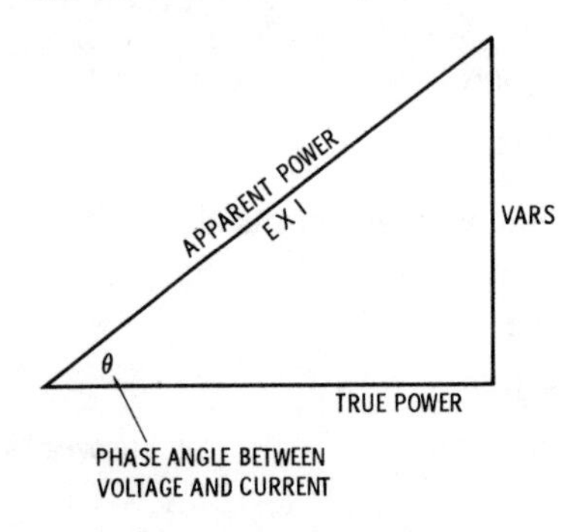

Fig. 6-6. The power triangle.

Q (Figure of Merit)

The Q of a coil is given by the equation:

$$Q = \frac{X_L}{R}$$

The higher the Q of a coil, the more sharply the resonant circuit containing that coil will tune.

The Q of a resonant tuned circuit is given by the equation:

$$Q = \frac{\text{Resonant Frequency}}{\text{Bandwidth}}$$

The bandwidth is measured between the points where the voltage has dropped to 70 percent of its maximum value, or the power to 50 percent of its maximum value.

The expression *figure of merit* is misleading because it is not always desirable for a circuit to tune sharply. For example, the tuned circuits in the i-f stages of a color receiver should have a response 4.2-MHz wide. In such circuits, a *swamping resistor* across the tuned circuit is used to lower its Q.

Resistors and Resistances

The resistance of resistors in series is given by the equation:

$$R_T = R_1 + R_2 + R_3 \ldots + R_n$$

In parallel, the circuit resistance can be found by computing the equivalent resistance of the resistors taken two at a time. The equation is:

$$R_{eq} = \frac{R_1R_2}{R_1 + R_2}$$

The reciprocal of resistance, called *conductance,* is represented by the letter *G.* The equation for conductance is:

$$G = \frac{1}{R} = \frac{I}{E}$$

When resistors are in parallel, their conductances are added to find the total conductance.

Resonant Frequency

The resonant frequency of a series-tuned circuit is given by the equation:

$$f_r = \frac{1}{2\pi\sqrt{LC}}$$

The resonant frequency of a series-tuned circuit is not affected by the presence of a series resistor, but the shape of the resonance curve becomes broader with increased resistance.

The resonant frequency of a parallel-tuned circuit is the same as for a series-tuned circuit, if the circuit resistance can be neglected. If there is resistance in the inductance or capacitance branches, then the resistance will affect the resonant frequency. The equation for parallel resonance with resistance in the branches is:

$$f_r = \frac{10^3}{2\pi\sqrt{LC}}\sqrt{\frac{R_L{}^2C - 10^6L}{R_c{}^2C - 10^6L}}$$

where,

f_r is the resonant frequency in hertz,
R is the resistance in ohms,
L is the inductance in henries,
C is the capacitance in microfarads.

This equation is important because it shows that a parallel LCR circuit can be tuned with a variable resistor. Since it is possible to have a negative value under the radical for certain combinations of R_c, R_L, L, and C, it follows that a given parallel circuit may not have a resonant frequency. This is different from a series RLC circuit which always has a resonant frequency.

Time Constant

The time constant of an RC combination is the amount of time required for the capacitor to charge through the resistor to 63 percent of the applied voltage. It is also the time required for the capacitor to discharge through the resistor to 37 percent of its original charged voltage. The time constant is given by the equation:

$$T = RC$$

Five time constants are required to fully charge or discharge a capacitor.

The time constant of an RL combination is the amount of time required for the current through the circuit to reach 63 percent of its maximum value when the circuit is energized, or drop to 37 percent of its maximum value when the circuit is de-energized. The time constant is given by the equation:

$$T = \frac{L}{R}$$

Transistor Parameters

In a common-base transistor circuit, the h_{FB} (dc alpha) is the ratio between the dc collector and emitter currents.

$$h_{FB}(\alpha) = \frac{I_C}{I_E}$$

where,

h_{FB} is the static value of the forward-current transfer ratio (common base),
I_C is the dc collector current,
I_E is the dc emitter current.

In a common-emitter transistor circuit, the h_{FE} (dc beta) is the ratio between the dc base and collector currents.

$$h_{FE}(\beta) = \frac{I_C}{I_B}$$

where,

h_{FE} is the static value of the forward-current transfer ratio (common emitter),
I_C is the dc collector current,
I_B is the dc base current.

In a common-base circuit, the h_{fb} (ac alpha) is the ratio of a change in collector current to the small change in emitter current that produced it.

$$h_{fb}(\alpha) = \frac{\Delta I_c}{\Delta I_e} \text{(collector voltage constant)}$$

where,

h_{fb} is the small-signal, short-circuit, forward-current transfer ratio (common base),
ΔI_c is the change in collector current (rms),
ΔI_e is the change in emitter current (rms).

In a common-emitter circuit, the h_{fe} (ac beta) is the ratio of a change in collector current to the small change in base current that produced it.

$$h_{fe}(\beta) = \frac{\Delta I_c}{\Delta I_b} \text{ (collector voltage constant)}$$

where,

h_{fe} is the small-signal, short-circuit, forward-current transfer ratio (common emitter),
ΔI_c is the change-collector current (rms),
ΔI_b is the change in base current (rms).

The ac alpha and the ac beta are usually referred to as simply alpha and beta. They are related by the equations:

$$\alpha = \frac{\beta}{\beta + 1}$$

and,

$$\beta = \frac{\alpha}{1 - \alpha}$$

The *cutoff frequency* of a transistor is the frequency at which α or β drops to 70.7 percent of its value at 1 kHz. The alpha cutoff and beta cutoff frequencies are an indication of how well the transistor will perform at high frequencies.

The *gain bandwidth* is the frequency at which the value of beta is unity (1.0). It is also an indication of how well a transistor will perform at high frequencies.

Tube Parameters

The three important tube parameters are:

$$\text{Amplification Factor } (\mu) = \left.\frac{\Delta E_p}{\Delta E_g}\right] \text{(plate current constant)}$$

$$\text{Plate Resistance } (r_p) = \left.\frac{\Delta E_p}{\Delta I_p}\right] \text{(grid voltage constant)}$$

$$\text{Transconductance } (g_m) = \left.\frac{\Delta I_p}{\Delta E_g}\right] \text{(plate voltage constant)}$$

These parameters are related by the equation:

$$g_m = \frac{\mu}{r_p}$$

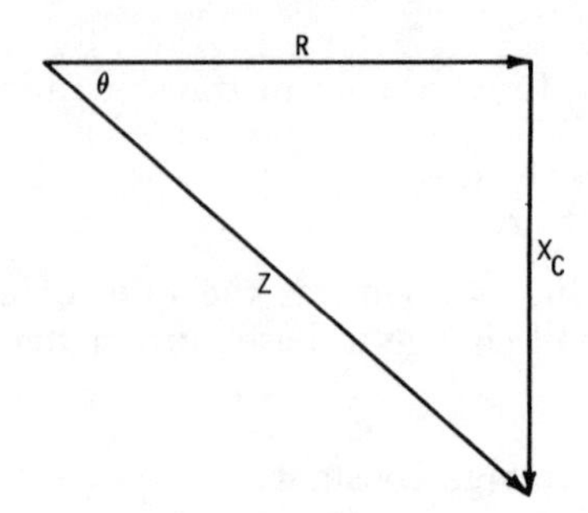

Fig. 6-7. The phase angle decreases as X_c becomes smaller.

Note that the amplification factor is obtained by dividing a voltage by a voltage. There are no units for amplification factor. Plate resistance, obtained by dividing a voltage by a current, is measured in ohms. Transconductance is obtained by dividing current by voltage. The unit of g_m is the mho, but micromhos are more frequently used as units.

PROGRAMMED QUESTIONS AND ANSWERS

Starting with question number 1, select the answer that you feel is correct. If you feel that (A) is correct, proceed to block number 17 as directed. If you feel that (B) is correct, proceed to block number 9 as directed. If you feel that more than one answer is correct, choose the one that you think is the *most* correct.

. . . .

1 To decrease the time constant of a series RL circuit,
(A) decrease the resistance. (Go to block number 17).
(B) decrease the inductance. (Go to block number 9.)

. . . .

2 The circuit is shown in Fig. 6-8.

$$I = \frac{E}{R}$$
$$= \frac{3}{30{,}000}$$
$$= 0.0001 \text{ ampere}$$
$$= 100 \text{ microamperes}$$

Here is the next question . . .
Three 30-ohm resistors are placed in parallel. The circuit resistance is:

Fig. 6-8. *Determine the current.*

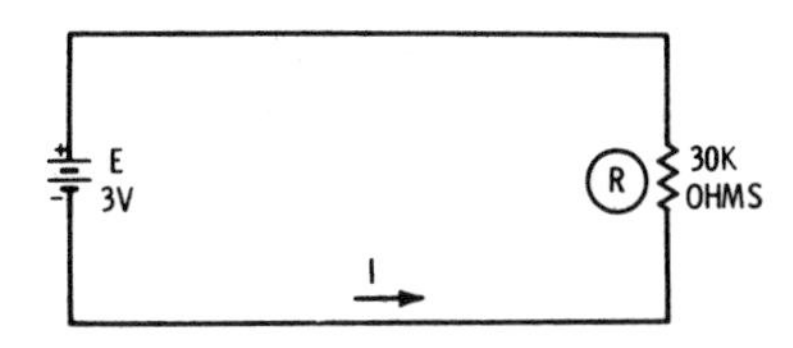

(A) 10 ohms. (Go to block number 6.)
(B) 90 ohms. (Go to block number 10.)

. . . .

3 Since the capacitor is not charged, the voltage across it is zero. This makes the voltage at *A* the same as the voltage at the positive terminal of the battery. Terminal *B* is connected to the negative terminal of the battery. Therefore, the full battery voltage appears across terminals *A* and *B*.

Here is the next question . . .

Which of the following will not increase the impedance of a series RC circuit?

(A) Increase the capacitance. (Go to block number 7.)
(B) Increase the resistance. (Go to block number 11.)

. . . .

4 Your answer is wrong. From the equation:

$$f_r = \frac{1}{2\pi \sqrt{LC}}$$

it is evident that increasing either the inductance or capacitance will reduce the resonant frequency. Go to block number 8.

. . . .

5 Your answer is wrong. Even though the resistance and inductance are in series, their impedance cannot be determined by simple addition. They are at an angle of 90°, and the impedance triangle must be used.

You can draw the triangle, letting one inch equal ten ohms.

Fig. 6-9. Resistance is doubled and the voltage remains unchanged.

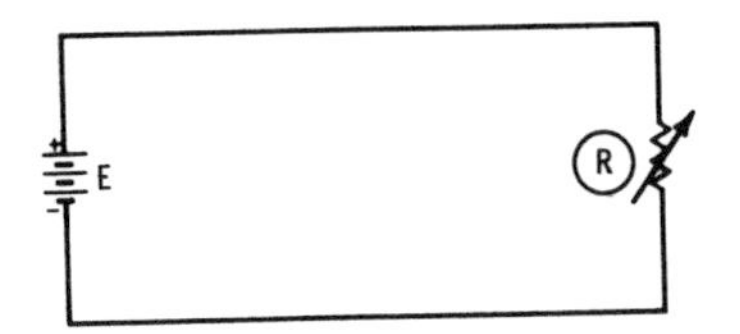

This will make the resistance line four inches long, and the inductive reactance line three inches long. (See Fig. 6-11.) The impedance line will be five inches long, which corresponds to a value of 50 ohms. Go to block number 23.

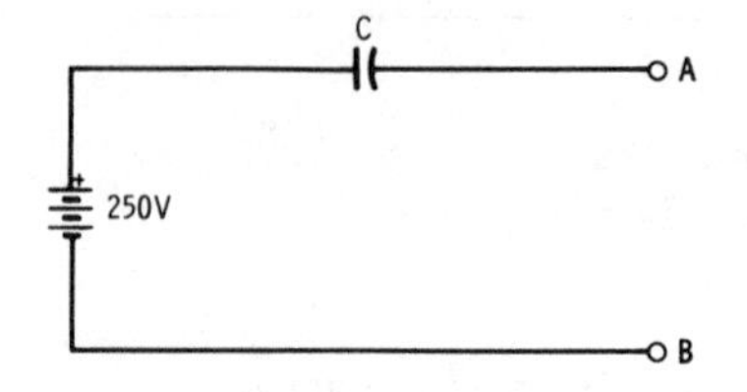

Fig. 6-10. Determine the voltage between A and B.

. . . .

6 When two resistors of equal value are in parallel, the combined resistance is one-half the resistance of either one. When three identical resistors are in parallel, the resistance is one-third; when four are in parallel, the resistance is one-fourth, and so on. The same is true of reactances and impedances.

Here is the next question . . .

In the circuit of Fig. 6-9 the resistance value is doubled. This will:

(A) increase the power dissipated. (Go to block number 14.)
(B) decrease the power dissipated. (Go to block number 18.)

. . . .

7 It is apparent from the impedance triangle (see Fig. 6-7) that decreasing the length of X_c will also decrease the length of Z, provided the length of R remains unchanged. Increasing the capacitance will reduce the value of X_c—that is, it will reduce the length of X_c on the impedance triangle.

Here is the next question . . .

Increasing the resistance in a series RLC circuit will:

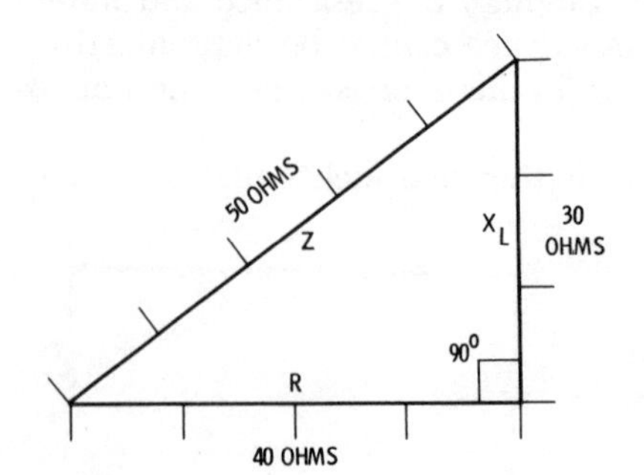

Fig. 6-11. Use 3-4-5 triangle to solve problem.

(A) increase the resonant frequency. (Go to block number 15.)
(B) not affect the resonant frequency. (Go to block number 19.)

. . . .

8 Decreasing either the inductance or the capacitance will increase the resonant frequency. This is evident from the equation:

$$f_r = \frac{1}{2\pi \sqrt{LC}}$$

Here is the next question . . .
A swamping resistor across a tuned circuit is used to:
(A) decrease the circuit Q. (Go to block number 12.)
(B) increase the circuit Q. (Go to block number 16.)

. . . .

9 Decreasing the inductance will decrease the time constant. This is evident from the equation:

$$T = \frac{L}{R}$$

Here is the next question . . .
In a series-tuned circuit, it is desired to increase the resonant frequency. This can be accomplished by:
(A) increasing the capacitance. (Go to block number 4.)
(B) decreasing the inductance. (Go to block number 8.)

. . . .

10 Your answer is wrong. Resistors in *series* add, but resistors in parallel combine by the reciprocal method. Go to block number 6.

. . . .

11 Your answer is wrong. Increasing the capacitance will decrease the capacitive reactance. This will decrease the impedance. Go to block number 7.

. . . .

12 Swamping resistors are used to reduce the Q and thus broaden the response of the circuit.
Here is the next question . . .
Increasing the capacitance in a series RC circuit will:
(A) decrease the power factor. (Go to block number 20.)
(B) increase the power factor. (Go to block number 24.)

. . . .

13 The larger voltage drop will be across the smaller capacitor. The actual voltage can be computed by the fact that the voltage is inversely proportional to the capacitance. See Fig. 6-12.

$$\frac{V_1}{V_2} = \frac{C_2}{C_1} = \frac{4}{1}$$

There is four times as much voltage across C_1 as there is across C_2. Divide the applied voltage into five parts (4 + 1). This gives a voltage of 20 volts across C_2 and a voltage of 80 volts across C_1.

Here is the next question . . .

An ac voltage has an average value of 100 volts. What is the rms value of this voltage?

(A) 111 volts. (Go to block number 26.)

(B) 63.6 volts. (Go to block number 25.)

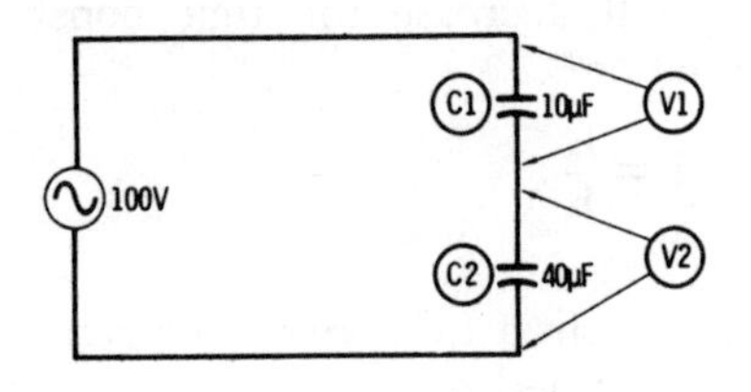

Fig. 6-12. Determine the voltage distribution.

. . . .

14 Your answer is wrong. Go to block number 18.

. . . .

15 Your answer is wrong. Go to block number 19.

. . . .

16 Your answer is wrong. Go to block number 12.

. . . .

17 Your answer is wrong. The equation for time constant is

$$T = \frac{L}{R}$$

From this equation you can see that decreasing the resistance will increase the time constant. This is true because decreasing the denominator of a fraction increases the value of the fraction. Go to block number 9.

. . . .

18 Power is equal to E^2/R. Since the voltage is constant, increasing R will decrease the power.

Here is the next question . . .

The capacitor in the circuit of Fig. 6-10 is uncharged. The voltage between terminals *A* and *B* is:

(A) zero volts because the capacitor cannot pass dc. (Go to block number 22.)

(B) 250 volts. (Go to block number 3.)

. . . .

19 For all practical purposes, there is no shift in frequency with a change in series resistance.

Here is the next question . . .

In a series RL circuit the resistance is 40 ohms, and the inductive reactance is 30 ohms. What is the impedance of the circuit?

(A) 50 ohms. (Go to block number 23.)

(B) 70 ohms. (Go to block number 5.)

. . . .

20 Your answer is wrong. Increasing the capacitance will decrease the capacitive reactance. Decreasing the capacitive reactance lowers the phase angle and increases the power factor. Go to block number 24.

. . . .

21 Your answer is wrong. The larger voltage drop will be across the smaller capacitor. Go to block number 13.

. . . .

22 Your answer is wrong. The key to the correct answer is the fact that the capacitor is not charged. Go to block number 3.

. . . .

23 If you remembered your 3-4-5 triangle, you noted that the impedance is 50 ohms by inspection.

Here is the next question . . .

A 10-microfarad capacitor is in series with a 40-microfarad capacitor across a 100-volt ac source. The larger voltage drop will be across:

(A) the 10-microfarad capacitor. (Go to block number 13.)
(B) the 40-microfarad capacitor. (Go to block number 21.)

. . . .

24 The impedance triangle of Fig. 6-7 shows why the power factor will increase. As X_c is made smaller (due to a larger capacitance), the phase angle (ϕ) becomes smaller. As the angle becomes smaller, its cosine becomes larger. Since the power factor is the cosine of the angle, it becomes larger as X_c becomes smaller.

If the capacitor is in parallel with the resistor, the opposite answer would be correct. In that case, increasing the capacitance decreases the reactance and causes a greater current through the capacitor. This makes the circuit look more capacitive to the generator, and the increased phase angle means a lower power factor.

The larger the power factor, the greater the useful power that is being consumed. The maximum value of power factor is 1.0, or 100 percent.

Here is the next question . . .

A 30K-ohm resistor is placed across a three-volt source. What is the current in microamperes? __________ μA. (Go to block number 2.)

. . . .

25 Your answer is wrong. Go to block number 26.

. . . .

26 The average value is 100 V. The peak value can be obtained by dividing the average value by 0.636.

$$\text{Average Value} = 0.636 \times \text{Peak Value}$$

Therefore,

$$\text{Peak Value} = \frac{\text{Average Value}}{0.636}$$

$$= \frac{100}{0.636}$$

$$= 157 \text{ volts}$$

The rms value can be found by multiplying 0.707 times the peak value.

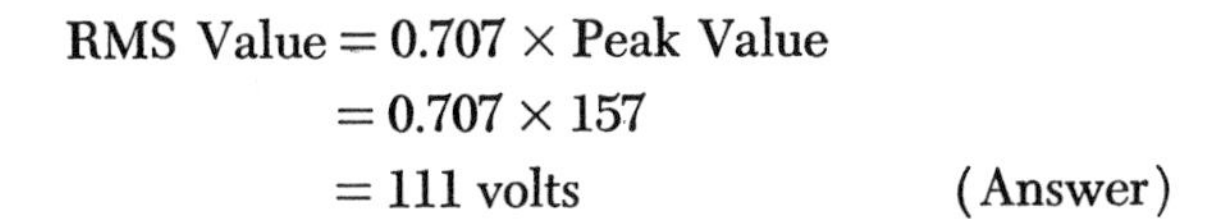

$$\text{RMS Value} = 0.707 \times \text{Peak Value}$$
$$= 0.707 \times 157$$
$$= 111 \text{ volts} \qquad \text{(Answer)}$$

You have now completed the programmed questions and answers.

. . . .

PRACTICE TEST

1. Two capacitors are placed in series across an ac voltage source. Each capacitor has a capacitive reactance of 10 ohms. The combined capacitive reactance of the two capacitors is:

(a) 5 ohms.
(b) 20 ohms.
(c) 100 ohms.
(d) 0 ohms.

2. Moving the plates of a capacitor closer together will:

(a) increase its capacitive reactance at a given frequency.
(b) decrease its capacitive reactance at a given frequency.
(c) not affect its capacitive reactance in a circuit provided the frequency is not changed.

3. In order to increase the resonant frequency of a series-tuned L-C circuit:

(a) decrease the inductance.
(b) increase the capacitance.
(c) lower its Q.
(d) reduce the phase angle.

4. A charge of 20 microcoulombs on a 10-microfarad capacitor will result in a voltage across the capacitor of:

(a) 200 volts.
(b) 0.2 volts.
(c) 2 volts.
(d) 20 volts.

5. In the circuit of Fig. 6-13, assume that the capacior is not charged. At the instant the switch is closed:

(a) the voltage across C will be maximum.
(b) the voltage across R will be maximum.

6. In the circuit of Fig. 6-13, the capacitor is considered to be fully charged:

(a) in a time period equal to R × C.
(b) in a time period equal to R/C.
(c) in a time period equal to C/R.
(d) in a time period equal to 5 × R × C.

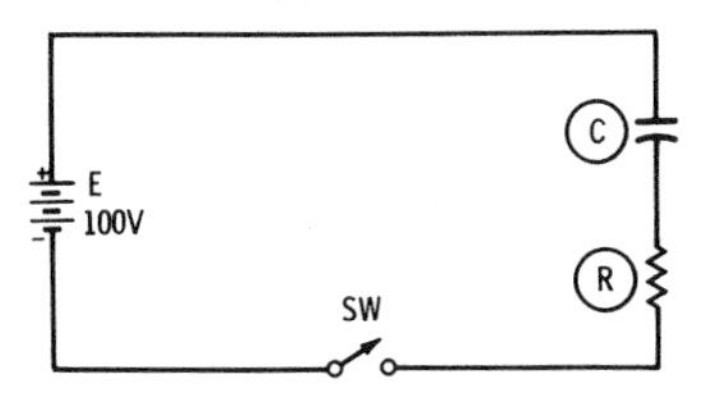

Fig. 6-13. Switch is open and capacitor is uncharged.

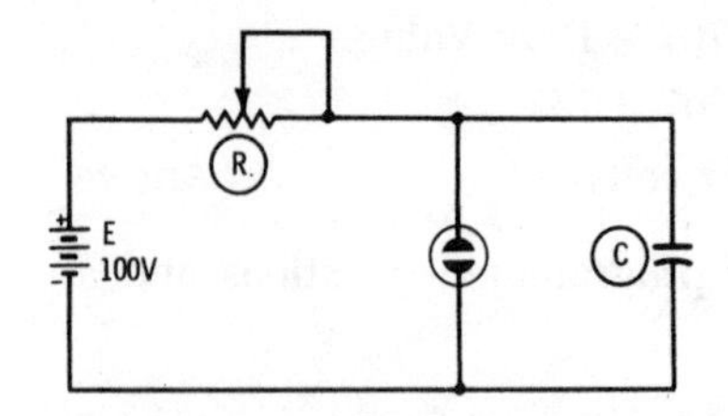

Fig. 6-14. A simple neon oscillator.

7. In the circuit of Fig. 6-13, moving the plates of the capacitor closer together will:

(a) not affect the time constant.
(b) increase the time constant.
(c) decrease the time constant.

8. In the simple neon oscillator circuit of Fig. 6-14, the neon lamp fires when the voltage across C reaches 60 volts. Each time the lamp fires, it discharges the capacitor. Which of the following statements is not true?

(a) Increasing the resistance of R will lower the frequency.
(b) Increasing the capacitance of C will lower the frequency.
(c) Increasing the voltage of E from 100 to 110 volts will decrease the frequency.

9. The circuit of Fig. 6-15 is:

(a) a differentiating circuit.
(b) an integrating circuit.

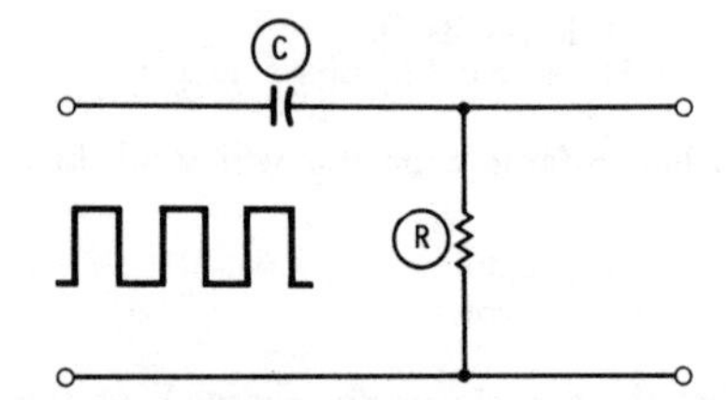

Fig. 6-15. A wave-shaping circuit.

10. Which of the following is the output waveform for the circuit of Fig. 6-15?

(a) The waveform in Fig. 6-16 will appear at the output of the circuit in Fig. 6-15.
(b) The waveform in Fig. 6-17 will appear at the output of the circuit in Fig. 6-15.

11. The time constant of a 500-picofarad capacitor in series with a 10-megohm resistor is:

(a) 5000 seconds.
(b) 50 microseconds.
(c) 5 milliseconds.
(d) 0.2 microseconds.

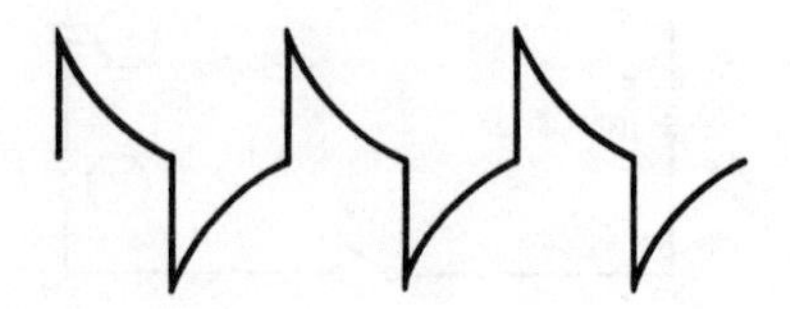

Fig. 6-16. Is this the output waveform for the circuit in Fig. 6-15?

Fig. 6-17. Is this the output waveform for the circuit in Fig. 6-15?

12. The capacitive reactance of a 100-pF capacitor at a frequency of one megahertz is:

(a) 1.59 ohms.
(b) 15.9 ohms.
(c) 159 ohms.
(d) 1590 ohms.

13. What is the resistance of five 100-ohm resistors in parallel?

(a) 20 ohms.
(b) 500 ohms.
(c) 10 ohms.
(d) 50 ohms.

14. A 500-ohm resistor is in series with a 300-ohm resistor. This series circuit is across a 40-volt source. The current flow is:

(a) 50 amperes.
(b) 50 microamperes.
(c) 5 milliamperes.
(d) 50 milliamperes.

15. To increase the time constant of a series R-L circuit:

(a) increase the resistance.
(b) increase the inductance.
(c) increase the frequency.

16. Two capacitors are in parallel across an a-c generator,

(a) their combined capacitance value is:

$$C_T = \frac{C_1 C_2}{C_1 + C_2}$$

(b) the capacitive reactance of the combination is the sum of their individual capacitive reactances.
(c) the capacitive reactance of the combination is less than the smaller reactance of the combination.
(d) the capacitive reactance of the combination is more than the larger reactance of the combination.

17. To increase the impedance of a series R-C circuit,

(a) increase the capacitance.
(b) increase the resistance.

18. In a series R-L circuit, the resistance and inductive reactance both have a value of 1000 ohms. The impedance is:

(a) 2000 ohms.
(b) 1000 ohms.
(c) 500 ohms.
(d) 1414 ohms.

19. For a series-tuned L-C circuit, increasing the inductance of the coil will:

(a) increase the resonant frequency of the circuit.
(b) decrease the resonant frequency of the circuit.

20. Which of the following statements is not true?

(a) A parallel-resonant circuit can be tuned by varying the resistance in one of the legs.
(b) A parallel-resonant circuit can be tuned by varying the inductance in one of the legs.
(c) A parallel-resonant circuit is sometimes called an *antiresonant circuit*.
(d) A series-tuned circuit can be tuned by varying the circuit resistance.

21. **A swamping resistor is placed across a tuned circuit to:**

(a) increase its resonant frequency.
(b) decrease its resonant frequency.
(c) lower its Q and broaden its frequency response.
(d) raise its Q and broaden its frequency response.

22. **A 40-μF capacitor is in series with a 60-μF capacitor across a 100-volt source. Which of the following statements is true?**

(a) There will be 60 volts across the 60-μF capacitor, and there will be 40 volts across the 40-μF capacitor.
(b) There will be 60 volts across the 40-μF capacitor, and there will be 40 volts across the 60-μF capacitor.
(c) The voltages across the two capacitors will be equal.

23. **A certain capacitor is charged to 100 volts. With the charge on the capacitor, the capacitor plates are moved further apart. Which of the following statements is true?**

(a) The voltage across the capacitor will decrease.
(b) There will be no change in the voltage across the capacitor.
(c) The voltage across the capacitor will increase.

24. **In the circuit of Fig. 6-18, the drain current is:**

(a) 10 mA.
(b) 1 mA.
(c) 100 microamperes.
(d) 10 microamperes.

25. **In the circuit of Fig. 6-18, the gate current is:**

(a) 0 mA.
(b) 1 mA.
(c) 10 mA.
(d) 100 mA.

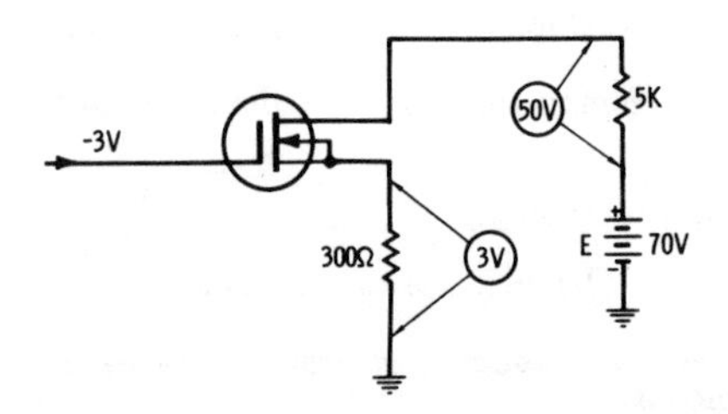

Fig. 6-18. A MOSFET circuit.

26. **In the circuit of Fig. 6-18, the voltage on the drain with respect to ground is:**

(a) 50 V.
(b) 3 V.
(c) —3 V.
(d) 20 V.

27. **A 30-millihenry inductor in a 10-kilohertz circuit has a dc resistance of 10 ohms. What is the Q of the coil?**

(a) 1884.
(b) 188.4.
(c) 18.84.
(d) None of these.

28. **The rms value of a sine-wave voltage is 100 V. What is the peak value?**

(a) 70.7 V.
(b) 141.4 V.
(c) 63.6 V.
(d) 111 V.

29. Which of the following is an indication of the upper-frequency limit of operation for a transistor?

(a) Alpha.
(b) Beta.
(c) Gain-bandwidth product.
(d) Leakage current.

30. Which of the following equations is correct?

(a) $G = \frac{E}{I}$.
(b) $T = \frac{R}{L}$.
(c) $g_m = \frac{\mu}{r_p}$.
(d) $\beta = \frac{I_c}{I_E}$.

31. There is exactly 10 volts across a resistor that is color coded brown, black, red, and gold. The maximum current that can be flowing through this resistor if it is within tolerance is:

(a) one microampere.
(b) 10,526 microamperes.
(c) 0.95 milliamperes.
(d) 0.0083 milliamperes.

32. Two capacitors are in parallel across a 1000-hertz generator. Each has a capacitive reactance of 50 ohms. Their combined capacitive reactance is:

(a) 100 ohms.
(b) 25 ohms.
(c) 40 ohms.
(d) 15 ohms.

33. A resistor that is color coded yellow, violet, red, and silver has a measured resistance value of exactly 5100 ohms. Is this resistor in tolerance?

(a) Yes.
(b) No.

34. In the circuit of Fig. 6-19, the capacitor is fully charged to 100 volts. At the instant the switch is closed,

(a) the voltage across the resistor will be maximum.
(b) the discharge current will be minimum.

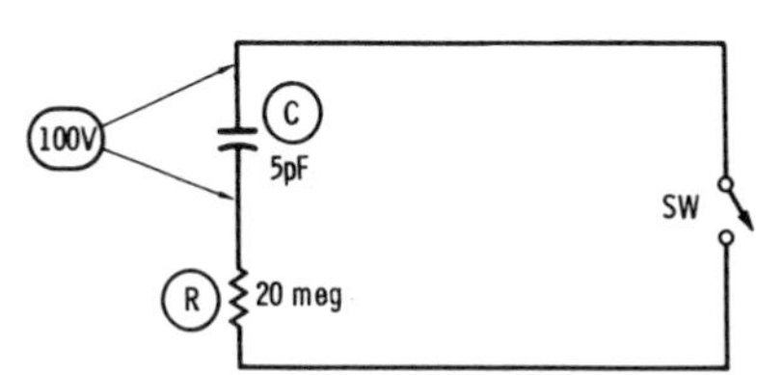

Fig. 6-19. A capacitive discharge circuit.

35. The time required for the capacitor in the circuit of Fig. 6-19 to discharge to 37 volts is:

(a) 67 microseconds.
(b) 37 microseconds.
(c) 10 microseconds.
(d) 100 microseconds.

36. The time required for the capacitor in the circuit of Fig. 6-19 to become fully discharged is:

(a) 335 microseconds.
(b) 185 microseconds.
(c) 50 microseconds.
(d) 500 microseconds.

37. **Which of the following statements is not true?**

(a) The Q of a given coil will be greater at 2 kHz than it is at 1 kHz.
(b) Neglecting any losses in the core, the Q of a coil with a high permeability core will be greater than the Q of an air-core with the same number of turns.
(c) The Q of a coil can be increased by spreading the turns of wire further apart.
(d) In a broadband circuit a coil with a high Q is not desirable.

38. **At 15 kHz a 1-millihenry coil has an inductive reactance of:**

(a) 943 ohms.
(b) 94.3 ohms.
(c) 9.43 ohms.
(d) 0.943 ohms.

39. **The capacitive reactance of a 100-pF capacitor at 15 kHz is:**

(a) 1,060,000 ohms.
(b) 106,000 ohms.
(c) 10,600 ohms.
(d) 1060 ohms.

40. **A certain capacitor is rated at 300 volts. Can this capacitor be used across a 230-volt ac line?**

(a) Yes.
(b) No.

41. **A 0.001-microfarad capacitor is placed in parallel with a 100-pF capacitor. The capacitance of the parallel combination is:**

(a) 100 pF.
(b) 1100 pF.
(c) 11 pF.
(d) 11,000 pF.

42. **A 3-pF capacitor in series with a 7-pF capacitor will give a combined capacitance of:**

(a) 10 pF.
(b) 4 pF.
(c) 2.4 pF.
(d) 2.1 pF.

43. **How much current will a 100-watt light bulb draw when placed across a 120-volt line?**

(a) 933 milliamperes.
(b) 833 milliamperes.
(c) 733 milliamperes.
(d) 633 milliamperes.

44. **A capacitor with a capacitive reactance of 10 ohms is in series with a 10-ohm resistor. The impedance of the circuit is:**

(a) 20 ohms.
(b) 5 ohms.
(c) 14.1 ohms.
(d) 12.8 ohms.

45. **A series circuit has the following reactance and resistance: $X_L = 250\Omega$, $X_c = 250\Omega$, $R = 250\Omega$. The impedance of this circuit is:**

(a) 250 ohms.
(b) 750 ohms.
(c) 83.3 ohms.
(d) 175.1 ohms.

46. **In a certain triode, a two-volt change in grid voltage produces a change in plate current of 10 mA. It takes a change in plate voltage of 30 volts to produce the same amount of plate current change. The amplification factor of the tube:**

(a) cannot be determined.
(b) is 60.
(c) is 15.
(d) is 3.

47. Which of the following transistor parameters always has a value of unity (that is, a value of one)?

(a) α.
(b) β.
(c) Gain-bandwidth product.
(d) None of these.

48. Which of the following circuits would have the maximum amount of impedance?

(a) A serie R-C circuit at a frequency whose period is equal to the time constant.
(b) A parallel R-L circuit at a frequency that causes $X_L = R$.
(c) A series L-C circuit at resonance.
(d) A parallel L-C circuit at resonance.

49. Three resistors in a parallel circuit each dissipate a power of 15 watts. The power dissipated by the three resistors in parallel is:

(a) 5 watts.
(b) 15 watts.
(c) 45 watts.
(d) 7.5 watts.

50. Three resistors in a series circuit each dissipate a power of 15 watts. The power dissipated by the three resistors in series is:

(a) 5 watts.
(b) 15 watts.
(c) 45 watts.
(d) 7.5 watts.

7

Monochrome Television Circuits

KEYED STUDY ASSIGNMENTS

Howard W. Sams Photofact Television Course
Chapter 12—RF Tuners
Chapter 13—Video I-F Amplifiers and Detectors
Chapter 14—Sound I-F Amplifiers and Audio Detectors
Chapter 15—Video Amplifiers
Howard W. Sams Color-TV Training Manual
Chapter 4—RF and I-F Circuits

The material covered by these keyed study assignments includes the principal circuits in monochrome receivers. It does *not*, however, cover sweep and high-voltage circuitry since they are discussed in a later chapter.

IMPORTANT CONCEPTS

This section deals with examples of circuitry used in transistorized television receivers.

Antenna Preamplifiers

In the earlier days of television it was not uncommon to employ a *booster* between the antenna and the input to the rf tuner of the television receiver. The booster was simply a wideband rf amplifier which was easily tuned for each channel. Its purpose was to increase the signal-to-noise ratio. The disadvantage of a booster is that it also amplifies all of the noise signals which enter through the antenna and transmission line system.

Better signal-to-noise ratio can be achieved by the use of an *antenna preamplifier*. This is simply a small, wideband rf amplifier mounted at the antenna terminals. An antenna preamplifier normally employs either common-emitter, or common-base, transistor amplifier circuits designed for very wideband use. Fig. 7-1 shows examples of antenna preamplifier circuits. Negative feedback (often called *degenerative feedback*) is used in the preamplifier circuit to reduce gain, and at the same time increase its bandwidth. Neutralization is used to offset the undesirable feedback through the base-to-collector junction capacitance.

The dc voltages for operating an antenna preamplifier are fed through the transmission line that connects the antenna with the television tuner. The dc power may come from the set, or it may be

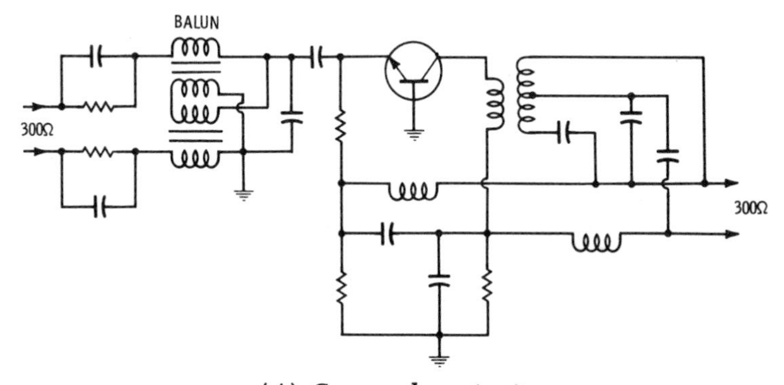

(A) Common-base circuit.

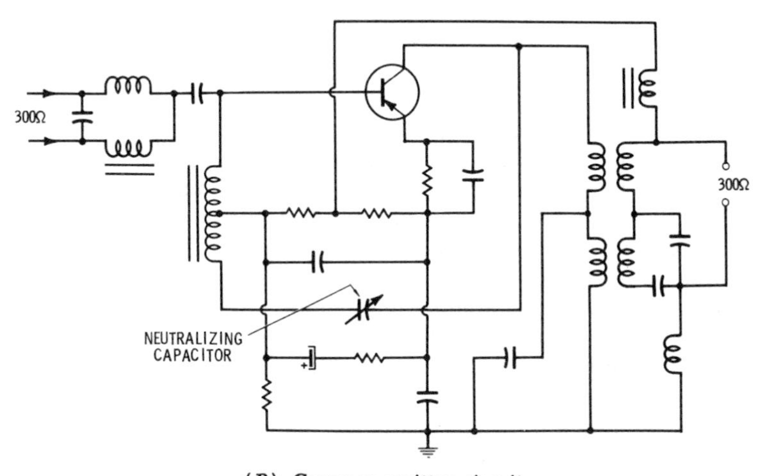

(B) Common-emitter circuit.

Fig. 7-1. Transistor preamplifier circuits.

from a separate dc (battery) or ac power supply located near the set.

Remote Controls

In monochrome receivers, the remote control system usually provides for *changing channels, adjusting volume,* and *turning the power on and off.* (For color-TV remote systems, adjustment for hue and color can also be accomplished.)

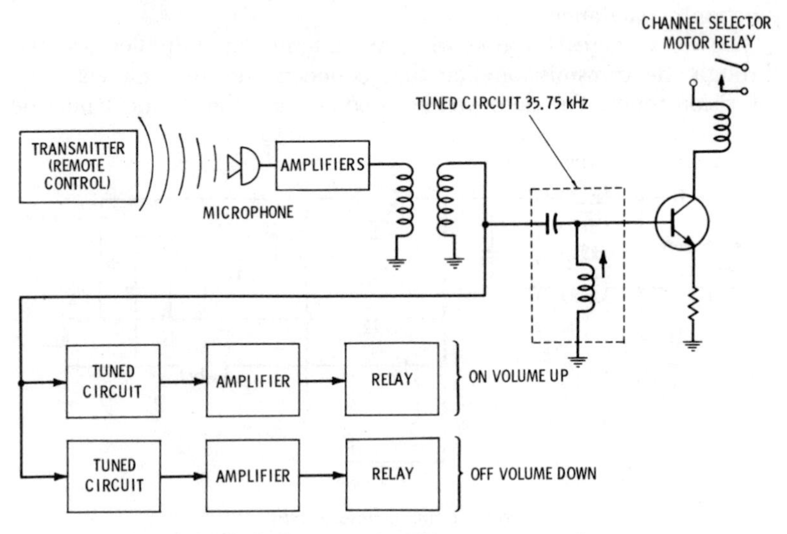

Fig. 7-2. Block diagram of a TV remote control system.

Fig. 7-2 shows a simplified diagram of a remote system for a monochrome TV receiver. The transmitter emits high-frequency sounds which are beyond the range of human hearing ability. These sounds have a different frequency for each function to be controlled. The microphone at the receiver picks up the sounds, amplifies them, and feeds them into a line from which each control circuit gets its signal.

Each control circuit is tuned to a different audio frequency. For example, the tuned circuit for the channel selector is tuned to 35.75 kilohertz. When this frequency is received, the tuned circuit will deliver a signal to the transistor which is normally cut off. The signal causes the transistor to conduct through a relay coil. Contacts on the relay energize a motor which turns the channel selector. Another frequency is used for volume control increase, and a third frequency is used for volume control decrease. As with many receivers, when the volume control is in a fully decreased position, the television receiver is switched off.

Transistor Tuners

Fig. 7-3 shows a simplified diagram of a transistor tuner. When the channel selector is changed from station to station, the following tuned circuits are altered: the antenna coils, the rf collector coils, the mixer base coils and the oscillator base coils.

The input signal to the transistor tuner is through a balun which converts the balanced transmission line input to an unbalanced configuration for operating the transistor amplifier. (If the transmission line used is unbalanced, as in the case of a coaxial cable, then the balun is bypassed. Some manufacturers make provisions for converting from balanced to unbalanced input.)

The antenna coils perform the job of preselection, and the rf collector coils tune the rf amplifier to the desired rf signal. Transistor rf amplifiers are normally neutralized, and this is indicated by the variable neutralizing capacitor shown in the circuit.

The rf signal is fed to a common-emitter mixer amplifier circuit. It is important to remember that the mixer stage in a television tuner, whether it is a transistor or vacuum-tube type, must be a nonlinear amplifier. Otherwise, heterodyning of the rf and oscillator signals will not occur. A neutralizing capacitor is shown in the mixer stage. Except for the fact that it is nonlinear, the mixer amplifier is like any other rf amplifier stage.

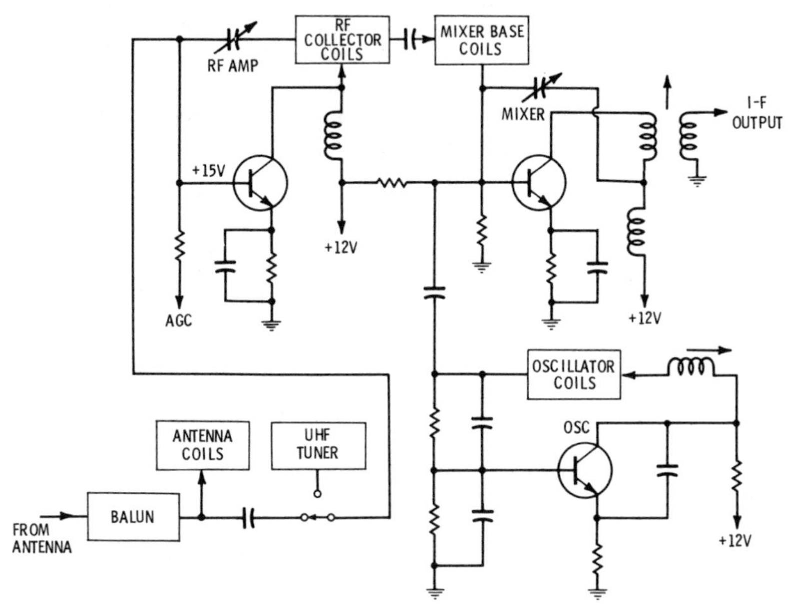

Fig. 7-3. Simplified schematic of a transistor tuner.

Transistors will readily oscillate in the uhf and vhf range due to collector-to-emitter feedback capacitance. Normally, an external capacitor is connected between the collector and emitter in order to enhance the feedback capacitance and assure oscillation. When you are analyzing a transistor oscillator circuit, you should always look for this feedback capacitor.

The uhf input to a transistor vhf tuner is normally through the rf amplifier stage. This input signal is an i-f signal from the uhf tuner. One important difference between the uhf tuner and the vhf tuner is that the uhf tuner seldom, if ever, has an rf amplifier. Also, the uhf mixer stage is usually a diode rather than a transistor amplifier.

A dual-gate MOSFET may be used for an rf amplifier or mixer. Fig. 7-4 shows a typical FET rf amplifier stage. The *dual-gate FET* has provisions for controlling the electron flow between the source and the drain by two electrodes—called gate No. 1 and gate No. 2. Gate No. 2 is supplied with the receiver agc voltage, and gate No. 1 receives the rf input signal from the antenna.

When the dual-gate FET is used as a mixer, the oscillator signal is applied to one of the gates and the amplified rf signal is applied to the other. Modulation or heterodyning takes place within the FET.

Some of the more recent innovations in transistor tuners is the use of voltage variable capacitors (sometimes called *varactors* or *varicaps*) as part of the frequency-determining tuned circuit. These varicaps are used in a number of different ways. In one application a different dc voltage is placed across the varicap for each channel being selected. This arrangement is sometimes used in a uhf tuner to provide step selection for the uhf channels in the area.

Another application of varactors is in automatic fine tuning (aft) circuits. Fig. 7-5 is a simplified diagram of an aft circuit. The pur-

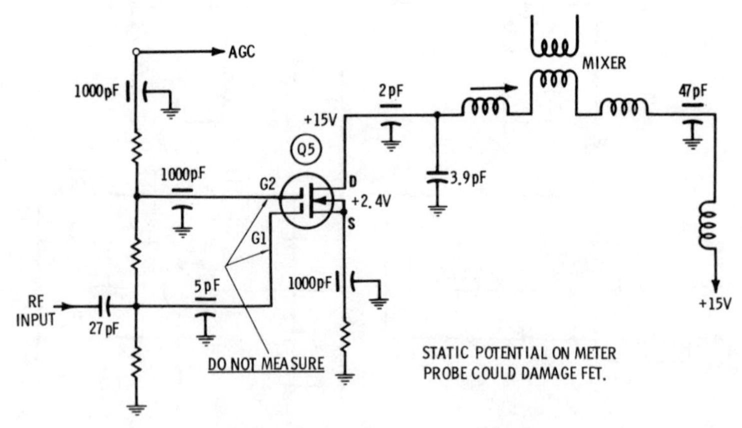

Fig. 7-4. Dual-gate rf amplifier.

pose of the aft circuit is to adjust the i-f picture-carrier frequency from the mixer to 45.75 megahertz. This is accomplished by the use of a feedback loop from the i-f amplifier section through a discriminator or ratio detector. The dc output voltage from the discriminator or detector is an *error correction* dc voltage. This voltage is applied to a varactor diode which is in the oscillator tuned circuit. When the oscillator frequency is exactly correct so that the video i-f carrier is at 45.75 MHz, then there is no correction voltage from the feedback loop.

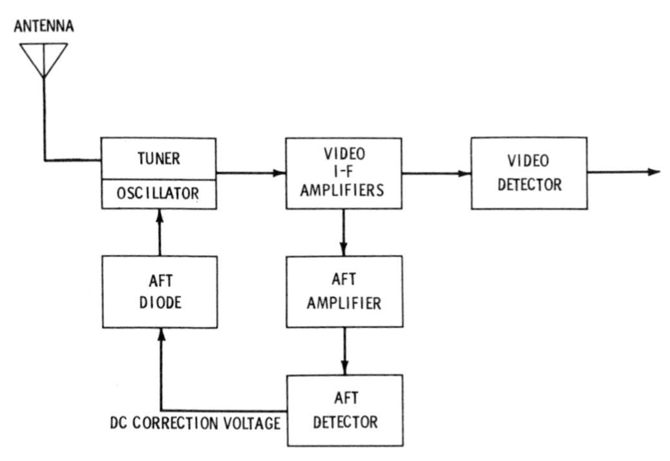

Fig. 7-5. The automatic fine tuning (aft) closed loop.

The use of an aft circuit is more important in color receivers than in monochrome receivers. This is because an improper adjustment of the manual fine-tuning control can result in a complete loss of the color signal. This has been the cause of many unnecessary color-television service calls.

When a receiver has an aft circuit, some provision is made for decoupling the circuit when the manual fine-tuning control is being adjusted. For example, the viewer must push the control to make the adjustment and this motion operates a switch that removes the aft correction voltage.

In the RCA circuit shown in Fig. 7-6, instead of using a varactor diode, the junction capacitance between the base and collector of a transistor is used. The emitter is not connected in this circuit.

I-F Amplifiers

Although the rf amplifier in a transistorized receiver may or may not be supplied with an agc voltage, the I-F amplifiers almost always are. There are two different ways to apply an agc voltage to a transistor amplifier. They are: *forward agc* and *reverse agc.* When reverse agc is employed, the gain of the amplifier is reduced by reducing the dc emitter current. This may be accomplished by the use of a dc control voltage which is applied to either the base or the emitter of the amplifier. With forward agc, a reduction in gain is achieved by controlling the collector voltage rather than the emitter current. In order to accomplish this, a resistor must be placed in the collector circuit. When the collector current increases, the collector voltage decreases due to the increase in voltage drop across this collector resistor.

To summarize, then, with reverse agc the emitter current is *reduced* to reduce the gain; and, with forward agc, the emitter current must be *increased* in order to increase the voltage drop across the collector resistor. This situation is somewhat different than that employed in most tube-type circuits in which the agc voltage normally reduces the plate current to reduce the gain of the amplifier stage.

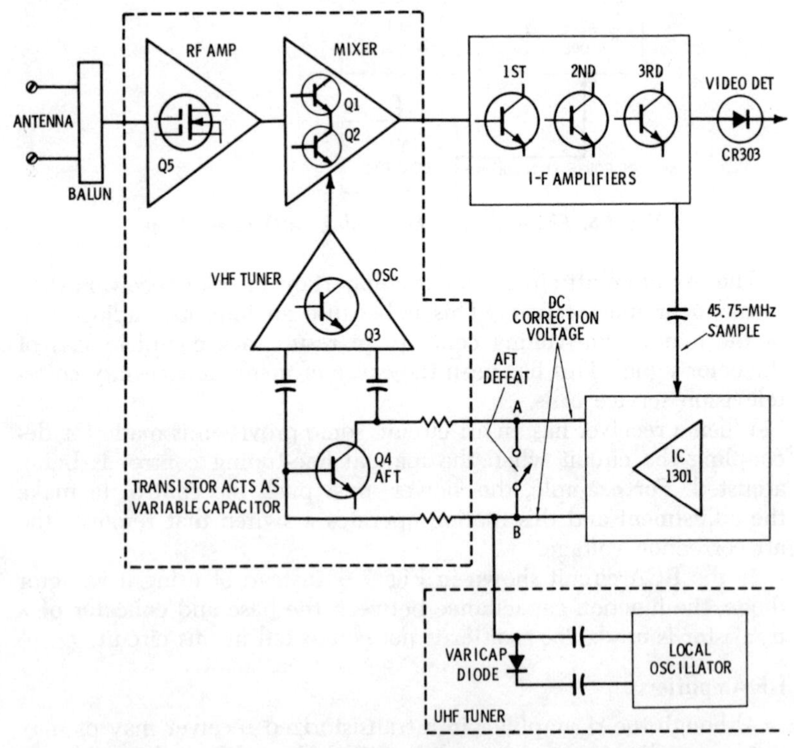

Fig. 7-6. Aft correction voltage applied to a transistor acting as a variable capacitor.

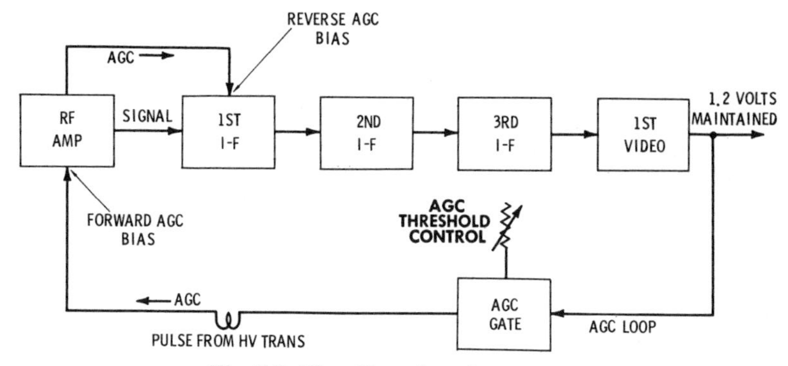

Fig. 7-7. Closed-loop keyed-agc system.

Sometimes a manufacturer will employ forward agc on one amplifier immediately followed by reverse agc on the next amplifier circuit. This is done in order to improve the bandpass characteristics of the amplifier.

The agc dc control voltage is obtained from a circuit that is similar to that used in tube-type receivers. Fig. 7-7 shows a typical transistor agc system. As in the case of the tube-type counterpart, the agc circuit is keyed into operation by a pulse from the flyback transformer. This makes the agc voltage dependent only upon the height of the sync pulses of the incoming rf signal, and independent of changes in the dc level due to changes in average lightness and darkness of the television scene.

Although not shown in Fig. 7-7, it is common practice to employ an agc amplifier to supply a greater amount of control voltage to the transistor amplifier circuits.

When analyzing amplifiers which are controlled by agc voltages, it is important to note whether or not the amplifiers are *stacked*—that is, arranged in such a way that the dc voltage for the collector of one amplifier is obtained from the emitter of the following amplifier. When amplifiers are stacked in this manner, an agc voltage applied to one will also affect the gain of another. (This is also true in amplifiers that are direct coupled.)

Detectors Used in Transistor-TV Receivers

Semiconductor diodes are used for the detector in transistor television receivers and their circuitry is very similar to that used in tube-type receivers. One difference that you may note is that the detector circuit may not be at dc ground potential. This is especially true if the detector is directly coupled to the first video amplifier.

An emitter-follower circuit may immediately follow the detector stage because the input impedance of the emitter follower is high

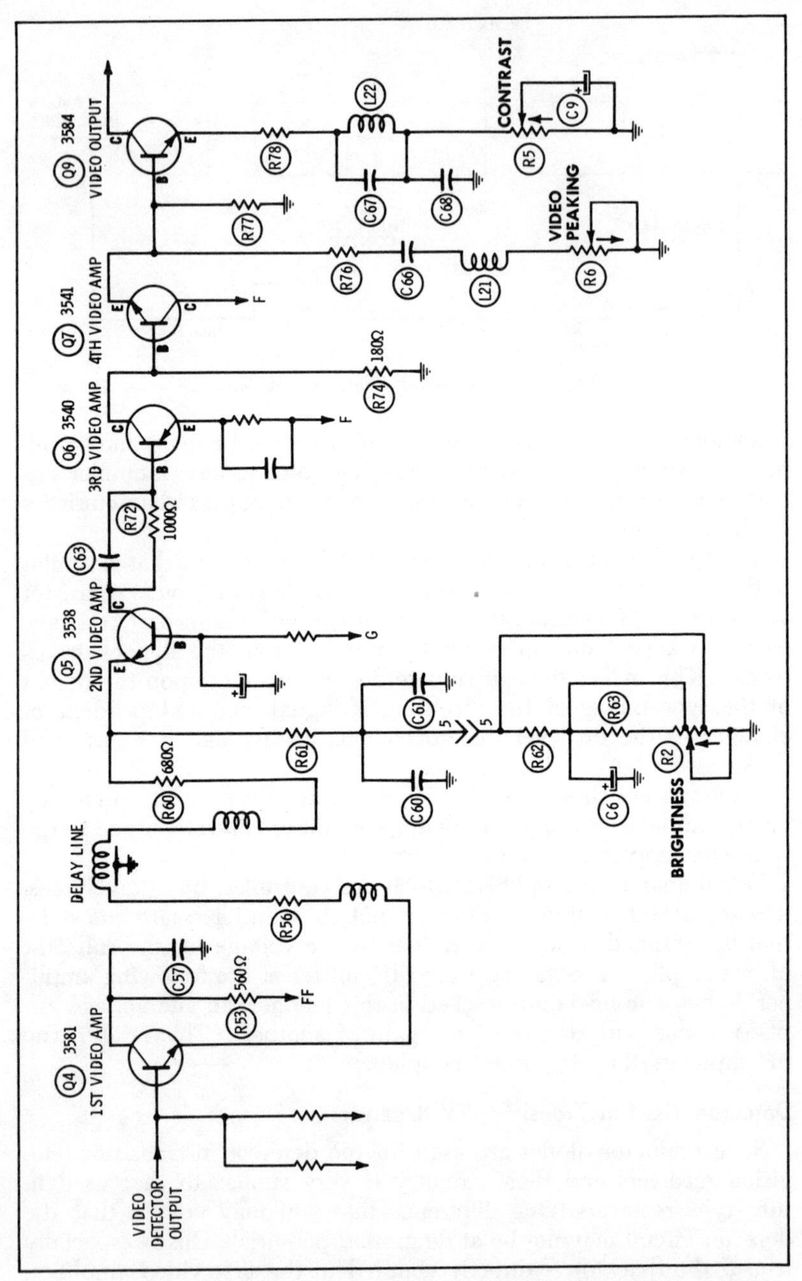

(A) *Brightness, video-peaking, and contrast controls.*

Fig. 7-8. Location of brightness, video-peaking,

and the output impedance is relatively low. This permits the detector to operate into a circuit which does not require any appreciable input power which, of course, the diode detector circuit could not supply.

Video Amplifiers

There are at least three controls that you may encounter in transistor video-amplifier circuits. When the transistor video amplifier is direct coupled to the picture tube, then the *brightness control* may be located in the video-amplifier circuit. The brightness control varies the amount of current through the video amplifier, and hence controls the amount of dc voltage on the picture-tube cathode.

The *contrast control,* which is also located in the video-amplifier stage, controls the amount of gain of the amplifier, or the amount of input signal to a video amplifier from a previous stage.

The *video-peaker control,* which is also called the *video optimizer,* is used to control the frequency response of the video amplifier. When the receiver is in a relatively noise-free signal area with a strong input signal, the video optimizer control is adjusted so that

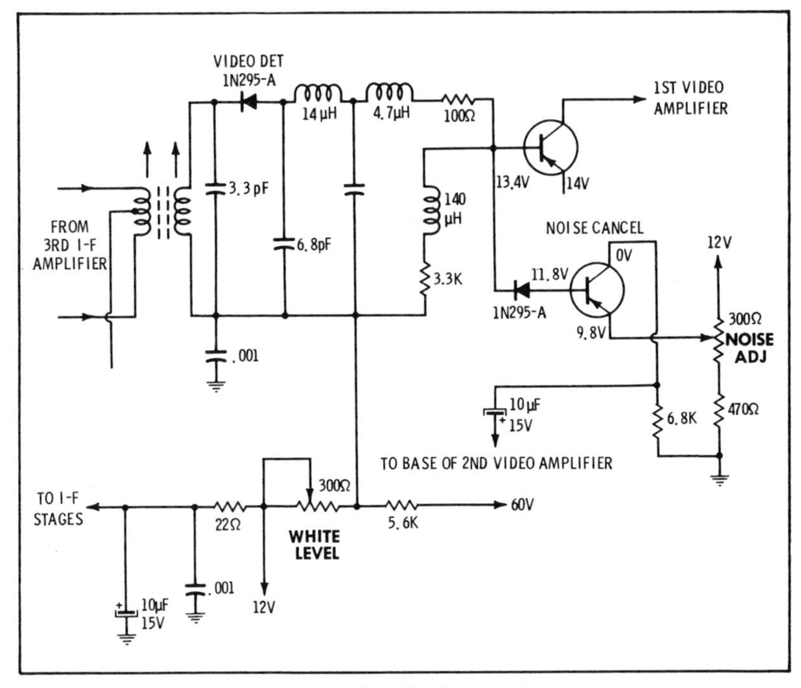

(B) White-level control.

contrast, and white-level controls.

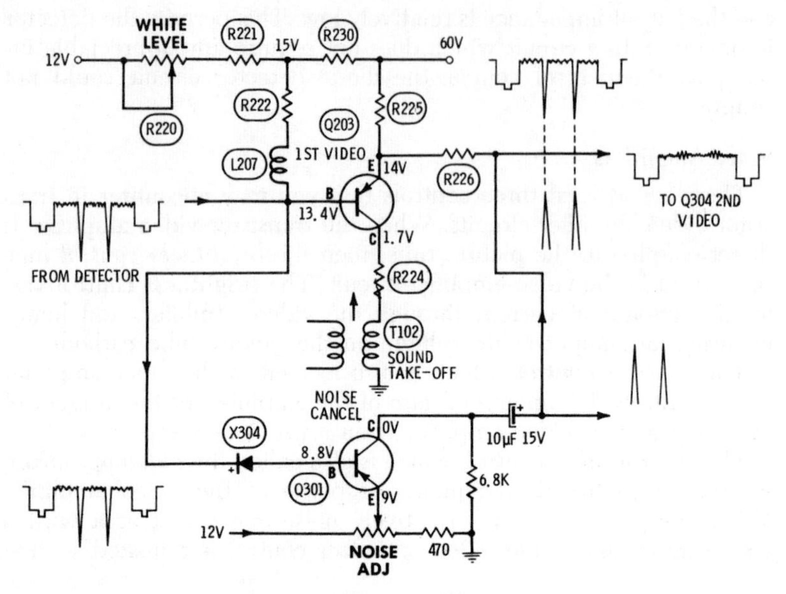

Fig. 7-9. Noise-cancelling circuit.

there is maximum video-amplifier frequency response. However, when there are noise signals present, the optimizer is adjusted to reduce the high-frequency response of the video amplifier, and thus reduce the effect of the noise.

A *white-level control* may be used to control the voltage on the base of the first video amplifier. This controls the gain of the stage, and determines the amount of signal amplitude delivered to the second video amplifier. The purpose of the white-level control is to set the level of the video signal near the base of the blanking pedestal, and thus obtain the maximum possible contrast range for the system.

The white-level control is normally set at the factory but may need to be readjusted if insufficient contrast range cannot be obtained in the receiver. Insufficient contrast range is, of course, also an indication of possible trouble in the agc or detector circuits. Fig. 7-8 shows some of the control circuits found in video-amplifier stages.

Noise-cancelling circuits may be used in the video-amplifier stage, but remember that the purpose of the noise-cancelling circuits is primarily to eliminate problems in synchronization. Noise spikes that occur in the video-signal range can trigger the vertical or horizontal sync oscillators and cause the picture to jump out of sync. To prevent this, a noise-cancelling circuit, such as the one shown in Fig. 7-9, may be used.

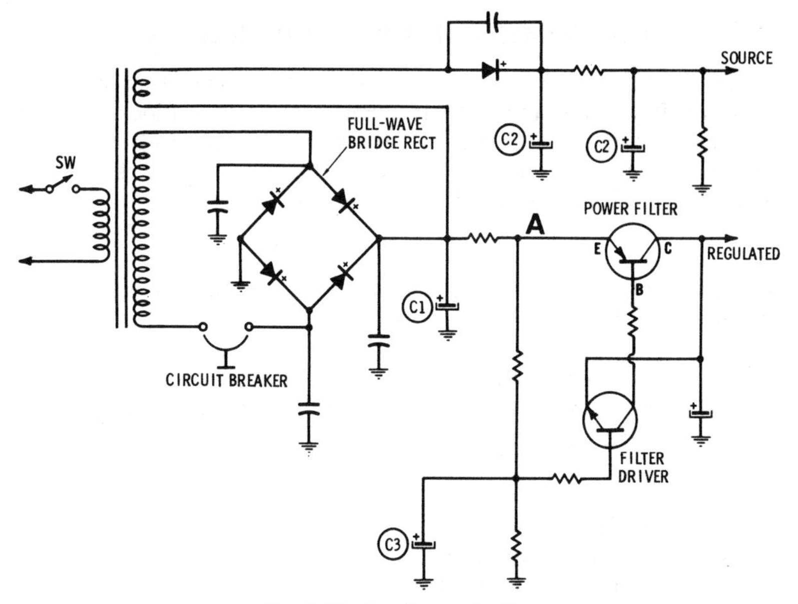

Fig. 7-10. An electronic filter.

Vertical and horizontal blanking signals may be applied to the video-amplifier stage or directly to the picture-tube grid or cathode in transistorized receivers.

Power Supplies for Transistor Receivers

In some transistor receivers—especially those which are operated by a battery or a low-voltage dc source—high voltages for B+ may be obtained from the flyback circuit. This method of obtaining B+ voltages is normally not used in tube-type receivers.

Another difference that you may notice in transistorized receivers is the use of regulated low-voltage power supplies. (This is especially true of transistorized color receivers.)

Fig. 7-10 shows a regulated dc power-supply circuit. This regulator can operate fast enough to remove ripples from the bridge rectifier, and therefore, it is sometimes referred to as a *power transistor filter* or an *electronic filter*. An increase in voltage at point A causes the filter driver to reduce the gain of the power filter, and conversely, a reduction in input voltage causes the filter driver to increase the gain of the power filter. Thus, the output at the regulated side is a constant voltage regardless of changes in voltage at point A.

Another method of obtaining a regulated voltage is to use a zener diode across the dc output voltage.

PROGRAMMED QUESTIONS AND ANSWERS

Starting with question number 1, select the answer that you feel is correct. If you feel that (A) is correct, proceed to block number 17 as directed. If you feel that (B) is correct, proceed to block number 9 as directed. If you feel that more than one answer is correct, choose the one that you think is the *most* correct.

. . . .

1 The dc voltage used for operating an antenna preamplifier is obtained by:

(A) a battery which is located in a sheltered compartment on the antenna mast. (Go to block number 17.)

(B) from a dc voltage or an ac voltage supply to the wires of a transmission line. (Go to block number 9.)

. . . .

2 collector and emitter.

Here is the next question . . .

Will a drift in the local-oscillator frequency of a tuner be more of a problem in receivers with intercarrier sound or split sound? (Go to block number 26.)

. . . .

3 Your answer is wrong. Go to block number 22.

. . . .

4 Your answer is wrong. When amplifiers are RC coupled, they are isolated as far as dc is concerned. Go to block number 11.

. . . .

5 Neutralization is frequently employed to cancel the feedback signal that goes through the base-collector junction capacitance. Incidentally, this capacitance is reduced by using a high collector-to-base dc voltage. Since this junction is reverse biased, a large voltage increases the depletion region and reduces the junction capacitance. The use of an epitaxial (nonconducting) layer permits a higher reverse voltage.

Here is the next question . . .

The type of agc used in transistor receivers in which the gain of the amplifier is decreased by reducing the collector voltage is called:

(A) forward agc. (Go to block number 24.)

(B) reverse agc. (Go to block number 18.)

. . . .

6 A MOSFET, like the one used for the rf amplifier, is easily destroyed by static charges. Manufacturers recommend that gate voltages should not be measured because the probe voltage can easily destroy the MOSFET. When they are stored, it is a good idea to short their leads together.

Here is the next question . . .

The mixer stage in a transistorized tuner must be:

(A) a linear amplifier. (Go to block number 25.)

(B) a nonlinear amplifier. (Go to block number 19.)

. . . .

7 Your answer is wrong. Emitter followers, sometimes called *common-collector circuits,* match a high impedance to a low impedance. Go to block number 13.

. . . .

8 Your answer is wrong. The agc circuit is a feedback loop between the detector and rf and i-f amplifiers. Control for this feedback is not likely to be in the video amplifier stage. Go to block number 15.

. . . .

9 The voltage for operating the transistor antenna preamplifier is fed through the transmission line. This may be either an ac or a dc voltage. If it is an ac voltage, a rectifier is located in the preamplifier case.

Here is the next question . . .

The remote control transmitter generates:

(A) a sound frequency above the human hearing capacity. (Go to block number 21.)

(B) an rf frequency in the channel-1 range. (Go to block number 14.)

. . . .

10 Your answer is wrong. Go to block number 22.

. . . .

11 Stacked amplifiers are dc coupled so that a change in current for one will change the amount of current in the other. The agc voltage in transistor amplifiers changes the gain by increasing or decreasing the dc current through the transistor.

Here is the next question . . .

Which of the following is a possible difference between the detector stage in a transistorized receiver and the detector stage in a tube-type receiver?

(A) Solid-state diode detectors are used in transistor receivers but not in tube-type receivers. (Go to block number 23.)

(B) The detector circuit may not be at dc ground potential in transistor receivers. (Go to block number 16.)

. . . .

12 Your answer is wrong. The agc voltage may affect the junction capacitance slightly by changing the size of the depletion region, but this is not significant as far as reducing the effect of signal feedback through that capacitance. Go to block number 5.

. . . .

13 A balun matches a BALanced line to an UNbalanced line.

Here is the next question . . .

The dual-gate FET used for an RF amplifier in an RCA tuner is an example of:

(A) a JFET. (Go to block number 20.)

(B) a MOSFET. (Go to block number 6.)

. . . .

14 Your answer is wrong. Go to block number 21.

. . . .

15 The brightness control *may* be located in the video-amplifier stage when that stage is direct coupled to the picture tube.

Here is the next question . . .

Regenerative feedback in the transistor tuner oscillator is usually between the ________ and the ________. (Go to block number 2.)

. . . .

16 The detector stage may not be at ground potential in transistor receivers.

Here is the next question . . .

The driver stage following the detector in a transistorized television receiver is likely to be:

(A) a common-emitter amplifier circuit. (Go to block number 10.)

(B) a common-base amplifier circuit. (Go to block number 3.)

(C) a common-collector amplifier circuit. (Go to block number 22.)

. . . .

17 Your answer is wrong. It would be very inconvenient to replace batteries on the mast. Go to block number 9.

. . . .

18 Your answer is wrong. Study the characteristics of forward and reverse agc again, then go to block number 24.

. . . .

19 The mixer stage must be nonlinear.
Here is the next question . . .
Undesired feedback of the signal through the base-collector junction capacitance is offset by the use of:
(A) keyed agc. (Go to block number 12.)
(B) neutralization. (Go to block number 5.)

. . . .

20 Your answer is wrong. The symbol as shown in Fig. 7-4 clearly shows that the gates are insulated from the channel. Go to block number 6.

. . . .

21 The sound generated by the transmitter is above the range of human hearing.
Here is the next question . . .
To convert the balanced input from a twin-lead transmission line to an unbalanced configuration, manufacturers supply:
(A) a balun. (Go to block number 13.)
(B) an emitter-follower circuit. (Go to block number 7.)

. . . .

22 A common-collector circuit, also known as an *emitter follower,* has a high input impedance which minimizes detector loading and a low output impedance.
Here is the next question . . .
Which of the following controls is most likely to be found in the video amplifier stage?
(A) The brightness control. (Go to block number 15.)
(B) The agc level control. (Go to block number 8.)

. . . .

23 Your answer is wrong. Solid-state diodes are often used in the detector stage of tube-type receivers. These diodes are sometimes called *crystal detectors.* Go to block number 16.

. . . .

24 Increasing the collector current, and thus reducing the collector voltage, reduces the gain of the amplifier when forward agc is employed.

Here is the next question . . .

In which of the following types of cascaded amplifiers will the gain of the second amplifier be affected by change in agc voltage on the first amplifier?

(A) Stacked amplifiers. (Go to block number 11.)
(B) RC coupled amplifiers. (Go to block number 4.)

. . . .

25 Your answer is wrong. Heterodyning will not take place in a linear amplifier. Go to block number 19.

. . . .

26 With a split sound receiver, a small drift in oscillator frequency can cause a complete loss of sound even though the picture is not affected.

You have now completed the programmed questions and answers.

. . . .

PRACTICE TEST

1. Noise-cancelling circuits are used in television receivers to:

(a) keep the noise out of the picture on the screen.
(b) prevent noise pulses from inadvertently triggering the sweep circuits.
(c) reduce static in the audio output.
(d) prevent degeneration in the 4.5-MHz sound detector stage.

2. The bandwidth of television tuner circuits is:

(a) 6 megahertz.
(b) 4.2 megahertz.
(c) 3.58 megahertz.
(d) 2.2 megahertz.

3. Which of the following is not a typical spurious response which a television tuner must be designed to reject?

(a) Adjacent channel sound carriers.
(b) Direct transmission through the rf system of signals in the intermediate-frequency range.
(c) Excess agc voltage.
(d) Overloading due to excessively strong signals from nearby stations.

4. Radiation of the local oscillator signal from a television receiver is prevented (or greatly reduced) by the use of:

(a) forward agc.
(b) an rf amplifier.
(c) twisting the twin-lead line from the antenna.
(d) reducing the dc voltage to a local oscillator.

5. An advantage of using a pentode over a triode for an rf amplifier is that:

(a) it produces a lower amount of noise.
(b) it is more linear.
(c) it does not require neutralization.
(d) it will operate with a lower power.

6. The selectivity and sensitivity of a tuner is governed primarily by:

(a) the rf amplifier stage.
(b) the mixer stage.
(c) the local oscillator stage.
(d) the method of selecting tuned circuits.

7. Triode rf amplifiers do not need to be neutralized when:

(a) tube shields are used.
(b) the amplifier is operated as a cathode follower.
(c) the amplifier is operated without an agc voltage.
(d) the amplifier is operated in the grounded-grid configuration.

8. A neutralized triode rf amplifier used in television tuners is:

(a) a cascode amplifier.
(b) a neutrode amplifier.
(c) a dynatrode amplifier.
(d) a slotted-line amplifier.

9. Which of the following is an advantage of cascode rf amplifier circuits?

(a) High signal-to-noise ratio.
(b) Low cost.
(c) Lower B+ voltage required.
(d) Neutralization is not required.

10. The video and sound carriers of a television signal are separated by:

(a) 3.58 megahertz.
(b) 4.5 megahertz.
(c) 21.9 megahertz.
(d) 45.75 megahertz.

11. Which of the following receivers is more sensitive to local oscillator drift?

(a) Split-sound receivers.
(b) Intercarrier receivers.

12. Which of the following is not an important factor in the choice of the intermediate frequency for a television receiver?

(a) Desired bandwidth.
(b) Selectivity.
(c) Image frequency.
(d) Whether tubes or transistors are used for the i-f amplifier.

13. In an intercarrier receiver, the ideal bandwidth of the i-f amplifier stage is:

(a) 5 megahertz.
(b) 2.5 megahertz.
(c) 1.8 megahertz.
(d) 4.5 kilohertz.

14. Fig. 7-11 shows an ideal i-f response curve for television receivers. The video carrier should be located at:

(a) point A.
(b) point B.
(c) point C.
(d) point D.

15. For the response curve of Fig. 7-11, the sound i-f frequency should be located at:

(a) point A.
(b) point B.
(c) point C.
(d) point D.

16. In order to prevent oscillation in the i-f stage as a result of undesirable regenerative feedback from one i-f amplifier to the other through a common power-supply connection, which of the following circuits is used?

(a) A 4.5-megahertz trap.
(b) A decoupling filter.
(c) A high-pass filter.
(d) Neutralization.

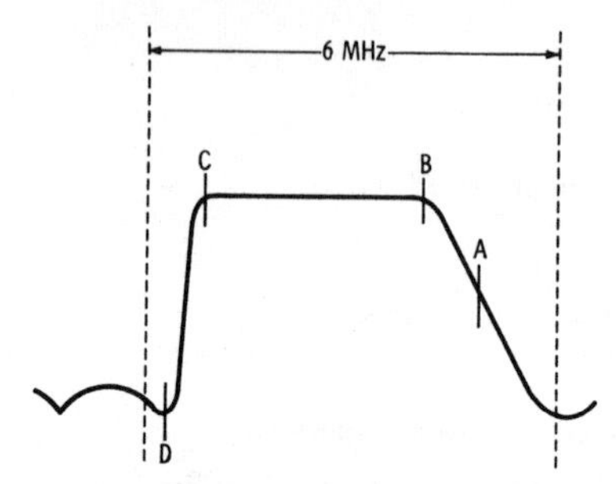

Fig. 7-11. Ideal i-f response curve.

17. Which of the following is a true statement?

(a) High-gain amplifiers are used when broadband amplification is required.
(b) Low-gain amplifiers are used when broadband amplification is required.

18. You are working on a battery-operated transistor portable which never worked from the time you took it out of the box from the factory. You note that the B+ line is coming from the flyback transformer. Which of the following is most likely to be correct?

(a) Someone at the factory incorrectly wired the receiver.
(b) This is a normal way to get a B+ voltage for some types of transistor television receivers.

19. Two video-amplifier stages are used between the detector stage and the picture tube of a certain television receiver. The video input signal to the picture tube is at the cathode. Which of the following statements is correct?

(a) The sync pulse at the output of the detector stage must be the most positive part of the signal.
(b) The sync pulse at the output of the video detector must be the most negative part of the signal.

20. Which of the following is not an advantage of using a semiconductor diode in a video-detector stage over using a vacuum-tube diode?

(a) The semiconductor diode has a lower dynamic resistance.
(b) The semiconductor diode does not require heated power.
(c) The semiconductor diode is easier to replace in the case of a malfunction.

21. An advantage of locating the second takeoff point at the plate of the video amplifier rather than at the grid is that:

(a) there is less chance of interference between the sound and video.
(b) there is the advantage of the additional gain supplied the video amplifier.

22. A circuit in which a single amplifier acts as both a sound i-f amplifier and an audio amplifier is called:

(a) a neutrode circuit.
(b) a reflex-amplifier circuit.
(c) a reflectodyne circuit.
(d) an impossibility.

23. Which of the following circuits is insensitive to amplitude modulation?

(a) A Foster-Seeley fm discriminator.
(b) A crystal-diode shunt-type video detector.
(c) A ratio detector.
(d) A slope detector.

24. The circuit of Fig. 7-12 shows:

(a) shunt-peaking compensation.
(b) series-peaking compensation.
(c) low-frequency compensation.
(d) transformer coupling.

25. The brightness control for a television receiver is not likely to be:

(a) in the filament circuit.
(b) in the video-amplifier transistor circuit.
(c) in the cathode circuit of the picture tube.
(d) in the grid circuit of the picture tube.

26. Which of the following components is likely to be associated with an automatic fine tuning control circuit?

(a) A voltage-dependent resistor.
(b) A varicap.
(c) A semiconductor controlled rectifier.
(d) A ferrite bead.

27. Which of the following circuits is not likely to be included in an integrated circuit?

(a) A sound i-f amplifier.
(b) An audio voltage amplifier.
(c) An fm detector circuit.
(d) An audio power amplifier.

28. In television rf amplifier circuits,

(a) tetrodes are never used.
(b) tetrodes must be neutralized.
(c) tetrodes can produce a high gain and relatively low noise comparable to cascode rf amplifiers.
(d) tetrodes are always used.

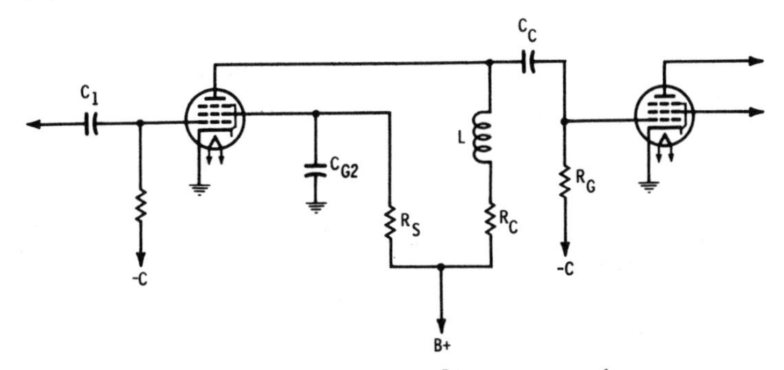

Fig. 7-12. A circuit with peaking compensation.

29. **A resistance across the tuned circuit of an i-f amplifier is used for:**

(a) increasing the gain of the stage and broadening the amplifier response.
(b) decreasing the resonant frequency of the tuned circuit.
(c) decreasing the gain of the stage and broadening the amplifier response.
(d) increasing the resonant frequency of the tuned circuit.

30. **Which of the following is the least popular type of tuner design?**

(a) Turret.
(b) Variable inductance.
(c) Disc.
(d) Switch type.

31. **For an intercarrier receiver, the sound takeoff point is most likely to be at:**

(a) the mixer stage.
(b) the i-f amplifier stage.
(c) the rf amplifier.
(d) the first video amplifier.

32. **In the i-f stages of a television receiver, traps are used for:**

(a) blocking rf signals.
(b) blocking sound signals.
(c) adjusting the shape of the i-f response curve.
(d) increasing the gain of the i-f amplifier.

33. **Which of the following is not a commonly used method of rejecting adjacent-channel carriers?**

(a) Parallel-tuned traps.
(b) Series-tuned traps.
(c) Cathode circuit degenerative traps in the i-f amplifiers.
(d) High-pass video-amplifier filters.

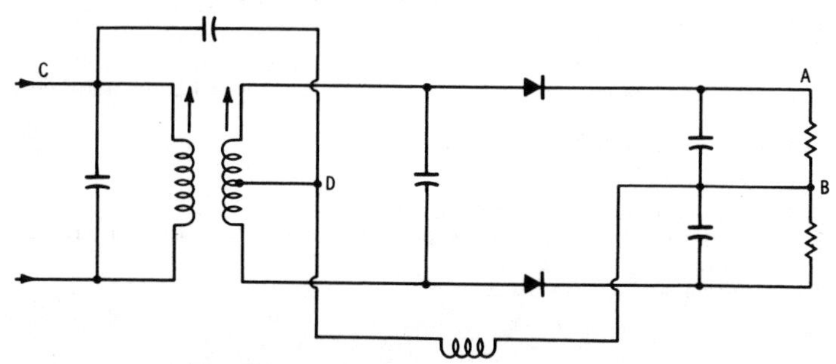

Fig. 7-13. A frequency-sensitive detector.

34. **The circuit of Fig. 7-13 is:**

(a) a ratio detector.
(b) a discriminator.
(c) a gated-beam detector.
(d) a slope detector.

35. **When used to demodulate an fm signal, the audio output signal is taken from the circuit of Fig. 7-13 at:**

(a) point A.
(b) point B.
(c) point C.
(d) point D.

36. **To obtain a better signal-to-noise ratio for an fm signal:**

(a) the high-frequency portion of the audio signal is de-emphasized in the modulation at the transmitter and a pre-emphasis circuit is used in the receiver.
(b) the high-frequency portion of the audio signal is pre-emphasized in the modulation at the transmitter and a de-emphasis circuit is used in the receiver.

37. In a fully-transistorized television receiver, which of the following types of detectors would not be used in the aft circuit?

(a) Gated-beam detector.
(b) Discriminator.
(c) Ratio detector.

38. Fig. 7-14 shows a:

(a) gated-beam detector.
(b) discriminator.
(c) ratio detector.
(d) slope detector.

39. In Fig. 7-14, the purpose of R_3 and C_4 is:

(a) decoupling.
(b) pre-emphasis.
(c) de-emphasis.
(d) scratch filter.

40. Which of the capacitors in the circuit of Fg. 7-14 is usually an electrolytic type?

(a) C_1.
(b) C_2.
(c) C_3.
(d) C_4.

41. A limiter circuit is associated with:

(a) discriminators.
(b) ratio detectors.
(c) gated-beam detectors.
(d) de-emphasis circuits.

42. A signal for vertical-retrace blanking comes from the vertical sweep circuit to:

(a) the video detector.
(b) the video i-f amplifier.
(c) the video amplifier.
(d) the agc circuit.

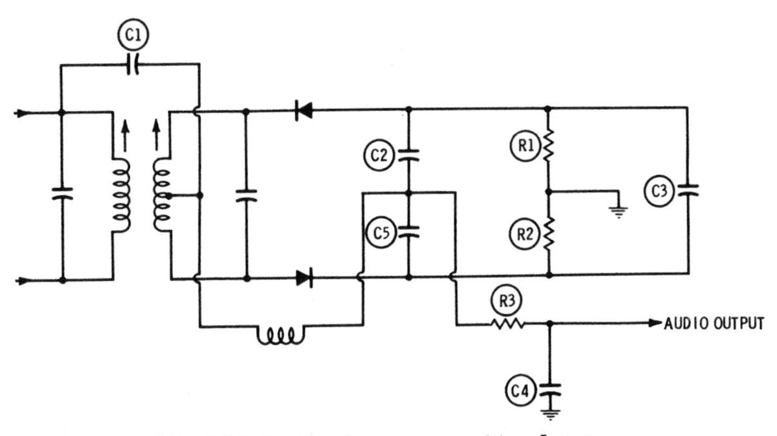

Fig. 7-14. Another frequency-sensitive detector.

43. The greatest portion of the total television-receiver gain is accomplished in:

(a) the rf amplifier.
(b) the i-f amplifier.
(c) the detector stage.
(d) the video amplifier.

44. The greatest bandwidth in a television receiver is in:

(a) the rf section.
(b) the i-f section.
(c) the detector stage.
(d) the video amplifier.

45. In a certain transistorized television receiver, the sync pulses are clipped off before the signal is applied to the video amplifier. Which of the following is true?

(a) This is done to eliminate buzzing noises in the sound section.
(b) This is done to prevent the sync pulses from getting into the picture.
(c) This is done to get better sync stability in the sync circuit.
(d) This is done to get a greater range of amplification from black to white in the amplified video signal for the purpose of driving the picture tube.

46. If the i-f amplifier in Fig. 7-15 is employing reverse agc, the gain of the amplifier will be reduced when the agc voltage is made:

(a) more positive.
(b) less positive.

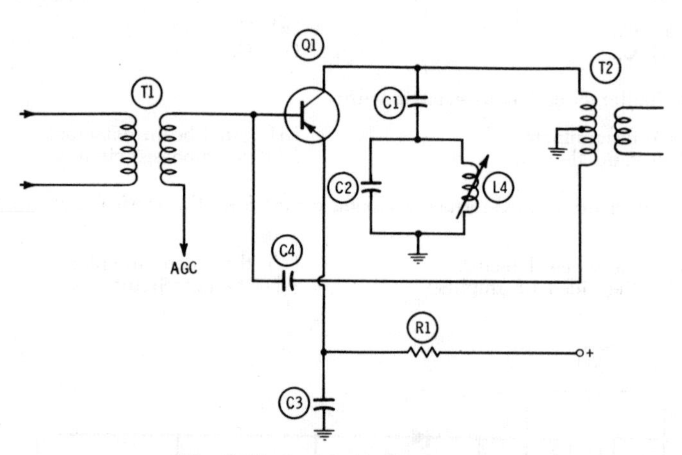

Fig. 7-15. An i-f amplifier circuit.

47. In the circuit of Fig. 7-15, components C2 and L4 are a:

(a) series-resonant trap.
(b) parallel-resonant trap.
(c) low-pass filter.
(d) high-pass filter.

48. Distortion in a diode detector circuit is most likely to occur when:

(a) the diode is operated at a relatively high power level.
(b) the diode is operated at a relatively low power level.

49. In a line-operated television receiver, a zener diode is used in the B+ power supply:

(a) for filtering.
(b) for rectifying.
(c) for voltage dividing.
(d) for voltage regulation.

50. You would expect to find a notch filter in:

(a) the power supply.
(b) the video amplifier stage.
(c) the i-f amplifier section.
(d) the sound section.

8

Color Television Circuits

KEYED STUDY ASSIGNMENTS

Howard W. Sams Color-TV Training Manual

Chapter 6—Bandpass-Amplifier, Color-Sync, and Color-Killer Circuits

Chapter 7—Color Demodulation

Chapter 8—The Matrix Section

IMPORTANT CONCEPTS

Knowledge of the control circuits in color receivers is helpful for a better understanding of the overall receiver system.

The controls described here are listed alphabetically. When several different names are given to the same control, they are cross referenced.

Agc Adjustment

The agc circuit in monochrome and color television systems is a closed-loop feedback circuit. In some circuits a variable resistor is used to control the amount of feedback, and therefore, set the maximum amount of gain obtainable in the rf and i-f amplifier stages. Fig. 8-1 shows a typical agc control circuit. Transistor Q1, the first video amplifier, is an emitter follower. A signal is taken from emitter resistor R3 and used to drive the base of Q2, the agc keyer. This is a *keyed agc system,* and Q2 will conduct only when a positive pulse arrives from the flyback-transformer winding. Thus, the agc system is not dependent upon the amount of video signal present; it is dependent only upon the strength of the sync pulses. The collector sig-

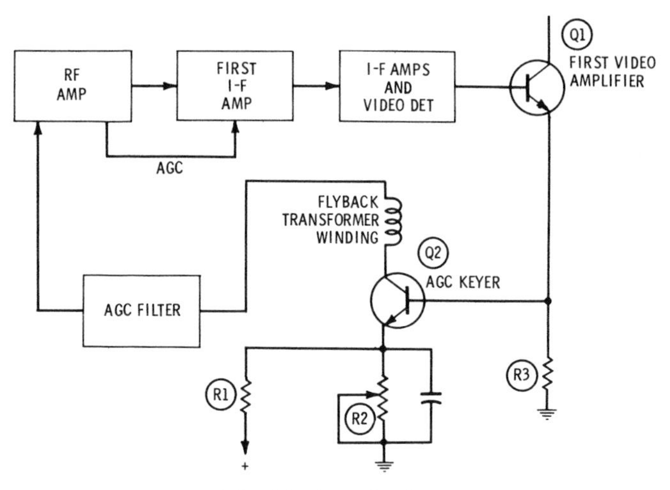

Fig. 8-1. Agc circuit.

nal from Q2 is fed to an agc filter which removes the pulsations. The output of the filter is a dc voltage used for regulating the gain of the rf and i-f amplifiers. A positive voltage is applied to the emitter of Q2 from voltage divider R1 and R2. This positive voltage holds the transistor at cutoff when there is no input signal present. Adjustment of R_2 establishes the amount of gain in Q_2, and thus establishes the level of agc voltage. This adjustment should be made for very strong input signals so that the strongest station to be received will not overdrive the video section.

Automatic Brightness Limiter Control

Some receivers employ a brightness limiter circuit to limit the amount of CRT beam current. The automatic brightness limiter (abl) control sets the maximum amount of beam current by limiting the amount of current through the abl transistor. This is not normally a customer adjustment. It is set by limiting the conduction through the abl limiter circuit to a specified current.

Automatic Color Control Adjustment

The automatic color control adjustment regulates the gain of the chroma amplifier circuit, and therefore, determines the maximum amount of color signal that can be delivered to the demodulators.

Background Control

The term *background* is used in reference to the luminance level of a raster. Background controls adjust the dc bias levels of the color-CRT guns in order to achieve the proper luminance level.

Beam Landing Control

Beam landing controls are used to insure that the beams from the three color guns of the picture tube arrive at the screen of the tube at exactly the right point. Proper beam landing is usually accomplished by adjustment of purity magnets and by positioning of the yoke.

Bias Control

The bias control is used for simultaneously adjusting the bias of the three guns in the color picture tube. It is used in conjunction with the background and screen controls to achieve the proper luminance level.

Brightness Control

The brightness control is a customer control that sets the amount of bias on the color picture tube. The brightness control is adjusted with the color control set at minimum. (The contrast control and the brightness control are always adjusted with a black-and-white picture.) The brightness control may be located in either the grid or cathode circuit of the picture tube. If the video amplifiers are direct coupled to the picture tube, the brightness control may be in the video amplifier circuit.

Brightness Range

This is not normally a customer adjustment. It is a variable resistor, usually in series with the brightness control, which sets the maximum amount of brightness on the picture tube. When properly adjusted, the brightness range prevents the customer from accidentally setting the brightness control too high.

Chroma Gain Control

This is another name for color control. (See Color Control.)

Chromatone Control

(See Tint Control.)

Cinema Control

(See Tint Control.)

Color Control

The color control adjusts the amount of color signal which is delivered to the color demodulators. This may be accomplished by adjusting the gain of the signal out of the chroma amplifiers (also known as the color-bandpass amplifiers) or it may be accomplished by adjusting the amount of available signal delivered by the chroma amplifiers. Fig. 8-2 shows one method for controlling the color.

The visual effect obtained by adjusting the color control is to change the saturation of the colors displayed. Varying the color control should not change the hue of the color. In other words, if the color control circuit is operating properly, adjusting the color control from its minimum to maximum range will not change the color of flesh tones. (If it *does* change the color of the flesh tones, there is a problem in the receiver, probably in the alignment of the chroma amplifier.)

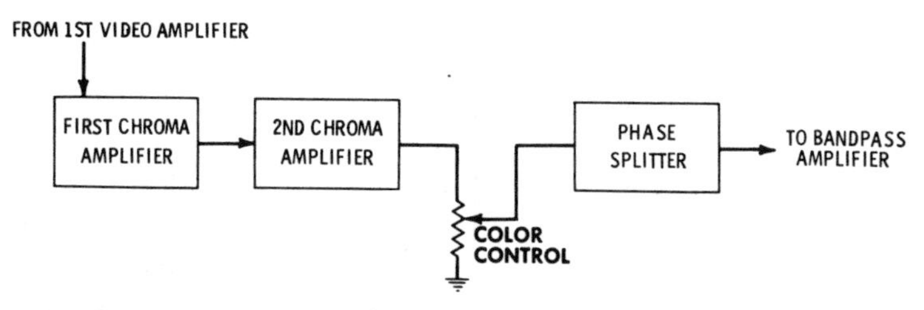

Fig. 8-2. Color control

Color Drive Control

The color drive control is a background control. Fig. 8-3 shows the dc operating controls that are related to the color picture tube. Note that the blue drive and green drive controls adjust the cathode voltages for the blue and green guns, while the bias control adjusts the voltage for all three guns. Individual screen controls determine the amount of electron beam that actually arrives at the face of the tube. These controls are interdependent, and they should be carefully adjusted according to manufacturer's specifications.

Colorfast Control

(See Tint Control.)

Color Gain Control

This is another name for color control. (See Color Control.)

Color Intensity Control

This is another name for color control. (See Color Control.)

Color Killer Control

The color killer control regulates the sensitivity of the color killer circuit so that it cannot be triggered by signals having less amplitude than the normal input signal. Fig. 8-4 shows two examples of color killer controls.

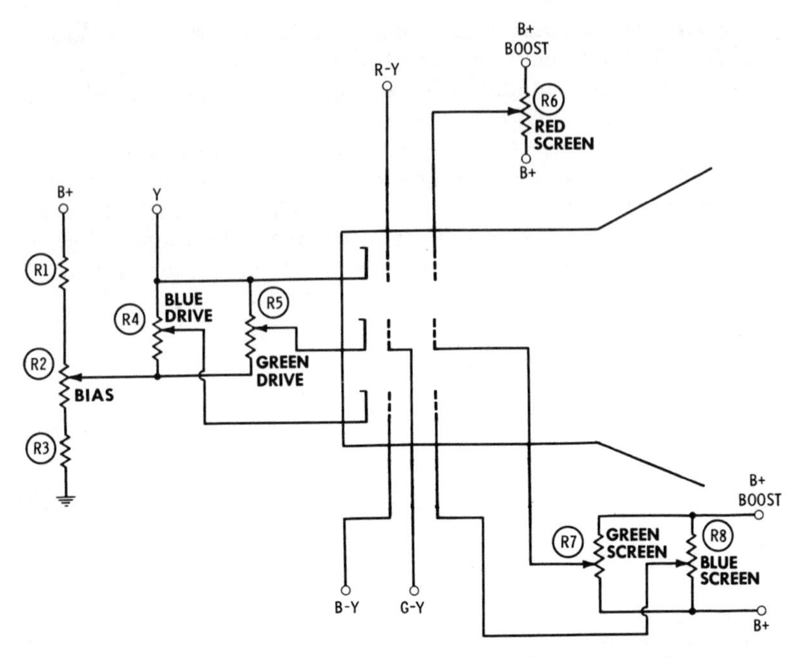

Fig. 8-3. Dc voltage-control circuits on color picture tube.

Color Phase Control

This is another name for hue control. (See Hue Control.)

Contrast Control

The contrast control sets the amount of luminance (or video) signal delivered to the picture tube. A contrast control is most easily adjusted with the color control turned off, that is, when a monochrome picture is being displayed. (Brightness and contrast are both adjusted with a monochrome picture displayed.)

Detail Control

This is another name for peaking control. (See Peaking Control.)

Fine Tuning Control

The fine tuning control varies the frequency of the local oscillator in the tuner over a limited range. Improper adjustment of the fine tuning control can result in a complete loss of color signals. When the fine tuning control is adjusted, the positions of the video carrier, color subcarrier, and sound carrier are established on the video i-f response curve. The fine tuning control is adjusted until maximum sound is obtained and an interference pattern between the sound

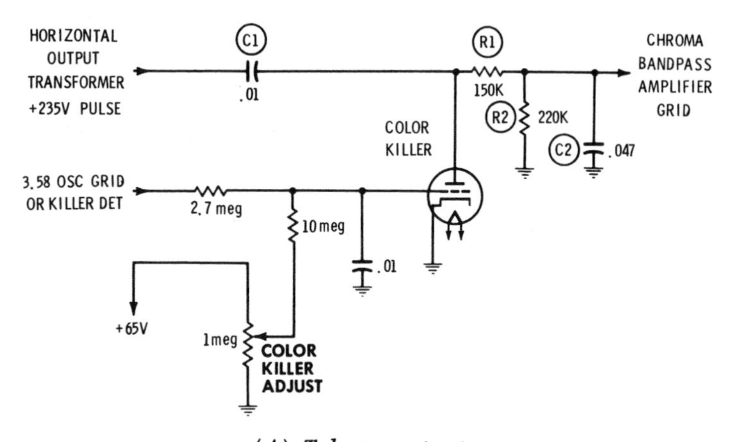

(A) *Tube-type circuit.*

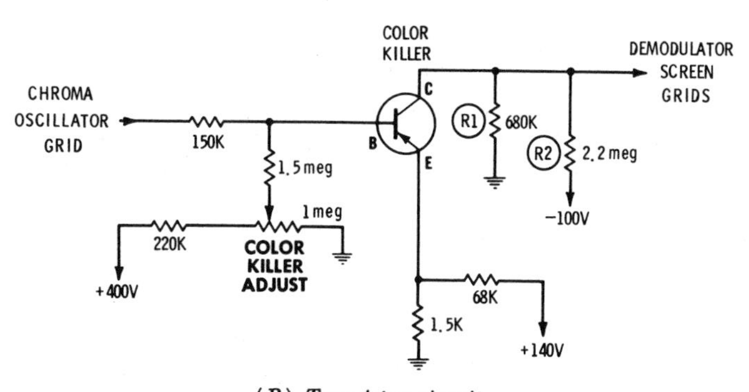

(B) *Transistor circuit.*

Fig. 8-4. Color killer circuits.

signal and the 3.58-MHz color subcarrier signal is observed on the screen. Then the fine tuning control is backed down until the interference pattern just disappears.

Because the fine tuning control is frequently misadjusted by the customer, some associated circuitry has been included in modern receivers to facilitate proper setting. One circuit is the *automatic fine tunning* (aft) circuit. This is a closed-loop system which is sometimes called the tuner automatic frequency control (afc).

Fine tuning indicators simplify the adjustment of the fine tuning control for the viewer. Fig. 8-5 shows an example of a fine tuning indicator (fti) circuit. A small amount of signal is sampled from the third video amplifier and applied to a 45.75-MHz tuned circuit. The output of this circuit is rectified by two diodes in a voltage-doubler configuration, and applied to Q1. Note that Q1, Q2, and Q3

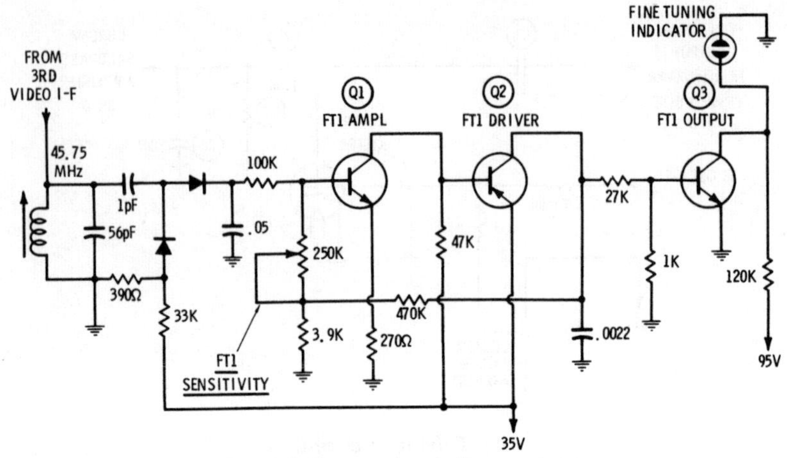

Fig. 8-5. Fine tuning indicator circuit.

are direct coupled. The dc voltage from the voltage doubler circuit saturates Q1, Q2, and Q3. With Q3 saturated the fine tuning indicator, which is a neon lamp, is grounded through the transistor. Therefore, it does *not* light when the frequency of the input is 45.75 MHz —that is, when the fine tuning control is properly adjusted. If the fine tuning is misadjusted, there is no input signal from the tuned circuit, and Q3 is no longer saturated. Instead, it acts as an open circuit. Under this condition the 95 volts across a 120K resistor causes the neon lamp to light, indicating to the viewer that the fine tuning control is not properly adjusted. Variations of this circuit include the use of a tuning-eye tube or a meter to indicate proper fine tuning adjustment.

FTI Sensitivity

The fti sensitivity control adjusts the level of input signal which will turn the neon lamp off in the circuit of Fig. 8-5. In other related circuits it adjusts the amount of dc voltage available on the tuning-eye tube or the dc range of the fti meter.

Hue Control

The hue control is also called the tint control in many receivers. (However, there is also another type of tint control used on Motorola receivers which should not be confused with the one that controls the hue of the displayed picture.) The hue control varies the *phase* of the 3.58-MHz color subcarrier as it is fed to the demodulator. Thus, the hue control can actually change the colors of the displayed picture. The hue control is adjusted for proper flesh tones. When the flesh tones are correct, then all of the other colors should also be correct in the picture. Fig. 8-6 shows a circuit with both a color control and a tint control (hue control) circuit.

Optimizer Control

This is another name for peaking control. (See Peaking Control.)

Peaking Control

The peaking control determines the high-frequency response of the video amplifier stage. Fig. 8-7 shows a typical circuit. It is easier to adjust the peaking control with the color control in the off position so that only a black-and-white picture is displayed. A peaking control is normally adjusted for maximum sharpness between the black and the white regions of the picture. This adjustment is made with a weak signal applied to the receiver. By limiting the high-frequency response of the video, it is possible to reduce some of the effects of high-frequency noise signals. It is a good idea to get the opinion of the customer on the adjustment of this control because some may prefer a sharp picture and others may prefer one that is *softer*. Remember that improper adjustment of the peaking controls can affect the colors, so always recheck the color picture after this control has been adjusted.

Phase Control

This is another name for the hue control. It gets its name from the fact that the hue control varies the phase of the 3.58-MHz signal. (See Hue Control.)

Pix Fidelity

This is another name for peaking control. (See Peaking Control.)

Sharpness

This is another name for peaking control. (See Peaking Control.)

Tint Control

On most receivers the tint control is the hue control which adjusts the phase of the 3.58-MHz signal delivered to the color demodulators. (See Hue Control.) However, there is another type of tint control which is also called "colorfast," "cinema," or "chromatone." This adjustment enables the customer to reduce the voltage on the green and blue guns in the color picture tube and thus produce a picture which is brown and white (instead of black and white). (The term "sepiatone" is often given to this hue in photography.)

Video Peaking Control

This is another name for peaking control. (See Peaking Control.)

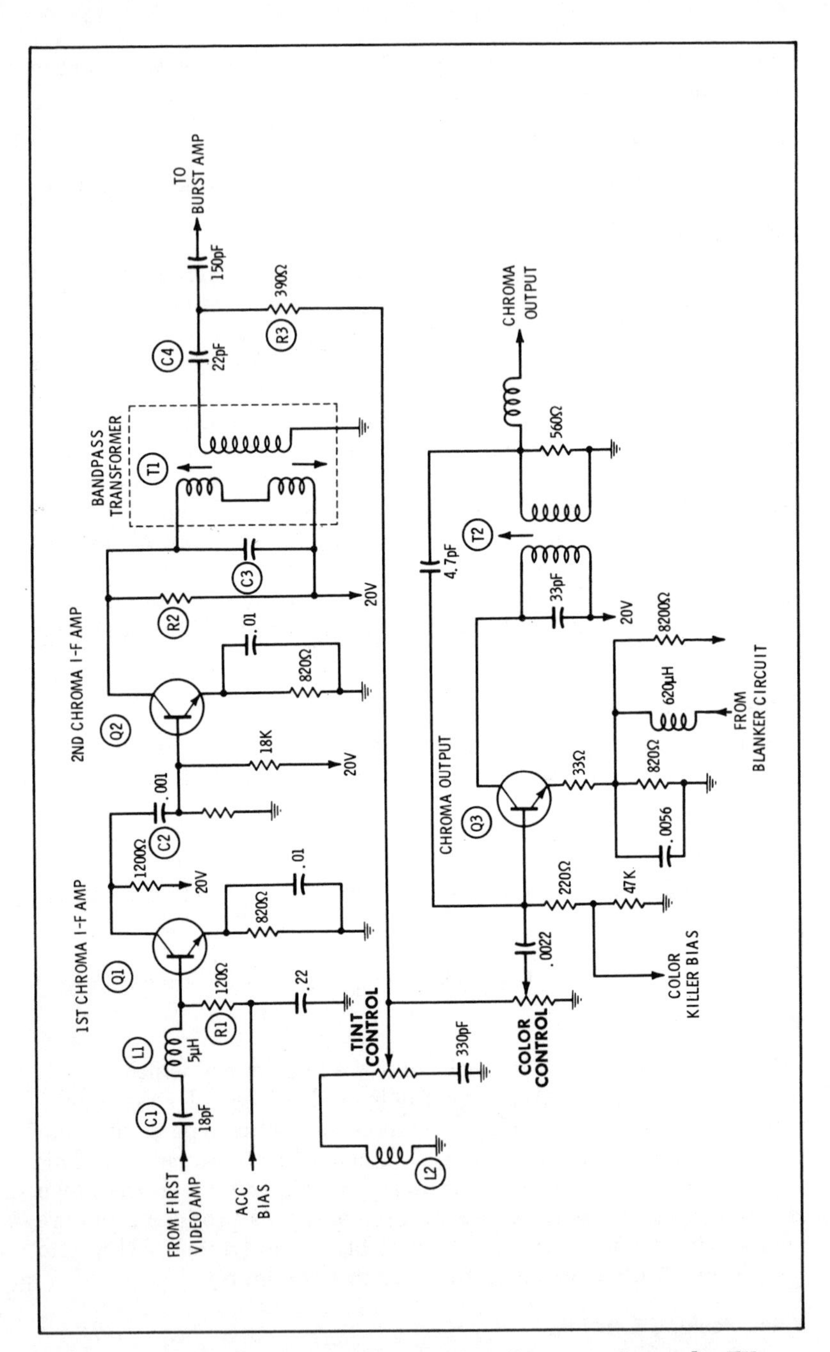

Fig. 8-6. Tint control and color control used in a transistor color TV.

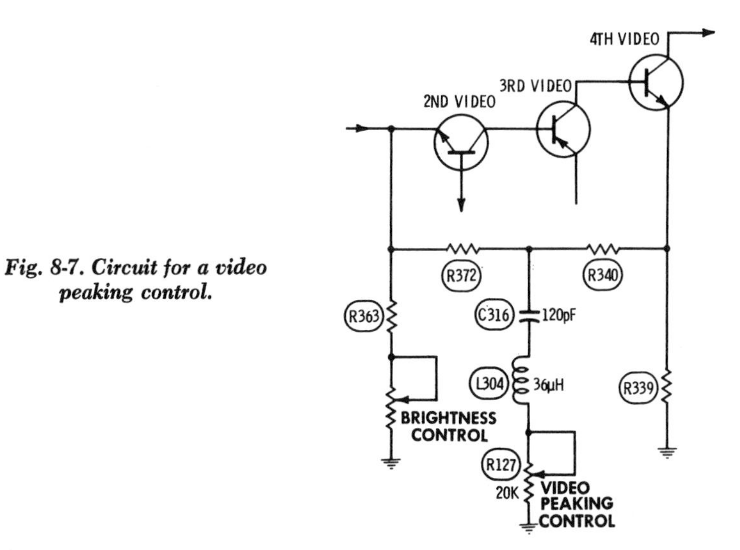

Fig. 8-7. Circuit for a video peaking control.

White Level Control

The white level control sets the amount of conduction of the picture tube. It is located in the video-amplifier circuit as shown in Fig. 8-8. It sets the range of video amplifier signal as delivered to the

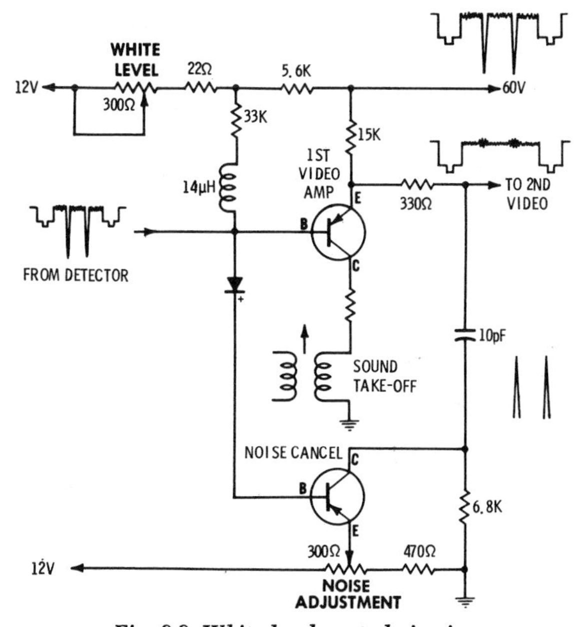

Fig. 8-8. White level control circuit.

picture tube.The white level control is not normally an adjustment available to the viewer. It may be necessary to readjust the white level control if a video amplifier transistor is replaced.

PROGRAMMED QUESTIONS AND ANSWERS

Starting with question number 1, select the answer that you feel is correct. If you feel that (A) is correct, proceed to block number 17 as directed. If you feel that (B) is correct, proceed to block number 9 as directed. If you feel that more than one answer is correct, choose the one that you think is the *most* correct.

. . . .

1 The hue control may also be called the:
(A) optimizer control. (Go to block number 17.)
(B) color phase control. (Go to block number 9.)

. . . .

2 Your answer is wrong. Changing the contrast control will change the *amplitude* of the luminance signal, but not the frequency response of the luminance channel. Go to block number 23.

. . . .

3 The automatic color control circuit is like an automatic gain control in the chroma circuits.
Here is your next question . . .
In a keyed agc system, the agc amplifier is keyed ON by a pulse from:
(A) the flyback transformer. (Go to block number 24.)
(B) the video amplifier. (Go to block number 19.)

. . . .

4 The tint control (hue control) should be adjusted for proper *flesh tones.*
Here is your next question . . .
Small changes in local-oscillator frequency are accomplished by varying the ________ control. (Go to block number 26.)

. . . .

5 Your answer is wrong. The optimizer control is in the video amplifier section. Go to block number 20.

. . . .

6 Your answer is wrong. The agc adjustment sets the receiver gain. Go to block number 12.

. . . .

7 Your answer is wrong. The automatic color control does not affect the amount of signal in the luminance channel. Go to block number 3.

. . . .

8 Your answer is wrong. The beam currents are never turned to maximum is normal operation of a color picture tube. Go to block number 10.

. . . .

9 The term "color phase control" is descriptive of the method by which the hue is changed since it varies the phase of the 3.58-MHz oscillator.

Here is your next question . . .

Beam landing is accomplished by adjusting the purity magnets and by:

(A) positioning of an ion trap. (Go to block number 25.)

(B) positioning of the yoke. (Go to block number 21.)

. . . .

10 Turning down the blue and green guns produces the reddish-brown colors that some viewers consider desirable.

Here is your next question . . .

The frequency response of the luminance channel is varied by adjusting the:

(A) contrast control. (Go to block number 2.)

(B) video peaking control. (Go to block number 23.)

. . . .

11 Your answer is wrong. Background is not the same as luminance. Go to block number 16.

. . . .

12 The automatic brightness limiter adjustment limits the beam current.

Here is your next question . . .

The tint control should be adjusted for proper ________.
(Go to block number 4.)

. . . .

13 Your answer is wrong. The tint control varies the phase of the 3.58-MHz signal, and therefore, changes the hue of the color. Go to block number 20.

. . . .

14 The agc control, sometimes called *agc bias,* is set with the receiver tuned to the strongest station.

Here is your next question . . .

The saturation of colors is changed by varying the:

(A) color control. (Go to block number 20.)

(B) tint control. (Go to block number 13.)

(C) optimizer control. (Go to block number 5.)

. . . .

15 Your answer is wrong. Turning down the red gun would leave blue and green colors in the picture. Blue and green do not combine to produce a brown color. Go to block number 10.

. . . .

16 The bias levels on the CRT are determined by the background controls.

Here is your next question . . .

The acc adjustment determines:

(A) the amount of video signal delivered to the luminance channel. (Go to block number 7.)

(B) the amount of signal delivered to the color demodulators. (Go to block number 3).

. . . .

17 Your answer is wrong. The optimizer control (which is often called the video peaker) sets the high-frequency response of the video amplifier. Go to block number 9.

. . . .

18 Your answer is wrong. The acc adjustment controls the amplitude of the color signal that is delivered to the demodulators. Go to block number 12.

. . . .

19 Your answer is wrong. The signal from the video amplifier contains all of the video information, including the noise spikes, that keying is supposed to eliminate. Go to block number 24.

. . . .

20 The color control should change the saturation level, but not the hue.

Here is your next question . . .

Sepiatone is a brown and white (rather than black and white) type of photograph. This type of picture is displayed on color receivers by:

(A) turning down the red gun. (Go to block number 15.)
(B) turning down the green and blue guns. (Go to block number 10.)
(C) setting all three guns for maximum beam current. (Go to block number 8.)

. . . .

21 Changing the position of the yoke will affect the adjustment of the beam landing.

Here is your next question . . .

A background control adjustment:

(A) sets the bias levels on the color CRT. (Go to block number 16.)
(B) sets the amplitude range of signal voltages in the luminance channel. (Go to block number 11.)

. . . .

22 Your answer is wrong. If you set the agc control with a weak signal, then a strong signal will overdrive the receiver. Go to block number 14.

. . . .

23 The video peaking control regulates the high-frequency response of the video amplifiers.

Here is your next question . . .

A control that sets the maximum beam current in the picture tube is the:

(A) acc adjustment. (Go to block number 18.)
(B) abl adjustment. (Go to block number 12.)
(C) agc adjustment. (Go to block number 6.)

. . . .

24 The keying pulse from the flyback transformer turns the agc amplifier on during the period of the sync pulse. Thus, any noise or changes in the average value of voltage in the video portion of the signal will not affect receiver gain.

Here is your next question . . .

The agc adjustment should be set with:

(A) a very strong signal. (Go to block number 14.)
(B) a very weak signal. (Go to block number 22.)

. . . .

25 Your answer is wrong. Ion traps were used with monochrome picture tubes to prevent destruction of the cathode by ionic bombardment. Go to block number 21.

. . . .

26 Small changes in local-oscillator frequency are accomplished by varying the *fine tuning* control.

You have now completed the programmed questions and answers.

. . . .

PRACTICE TEST

1. In a television receiver, the local-oscillator frequency can be varied by changing the:

(a) acc adjustment.
(b) abl adjustment.
(c) agc adjustment.
(d) fine tuning control.

2. The maximum possible gain of the video i-f section is set by the:

(a) agc adjustment.
(b) afc adjustment.
(c) acc adjustment.
(d) abl adjustment.

3. If the flesh tones change when the color control is varied, this is:

(a) an indication that the circuit is working properly.
(b) an indication that the video i-f section may need to be aligned.
(c) an indication that the chroma circuit may need alignment.
(d) an indication that the delay line is open.

4. The phase of the color subcarrier signal is varied by turning the:

(a) color control.
(b) tint control.
(c) optimizer control.
(d) agc control.

5. The aft correction voltage alters the:

(a) local-oscillator frequency.
(b) chroma-amplifier bandpass.
(c) maximum CRT beam current.
(d) luminance delay time.

6. A certain color receiver displays color but no picture. Which of the following is the most likely cause?

(a) Fine-tuning control improperly adjusted.
(b) Agc control not properly adjusted.
(c) Color-killer circuit not functioning properly.
(d) Open delay line.

7. If the beam landing controls are not properly adjusted, it will affect:

(a) only the monochrome picture.
(b) both the monochrome and the color picture.
(c) only the color picture.
(d) neither the monochrome nor the color picture.

8. The circuit that automatically disables the chrominance channel when no color signal is being received is the:

(a) acc circuit.
(b) abl circuit.
(c) color-killer circuit.
(d) luminance circuit.

9. A circuit that separates the chrominance signal from the composite color signal, and feeds it to the demodulators is the:

(a) color killer.
(b) luminance amplifier.
(c) burst amplifier.
(d) bandpass amplifier.

10. The composite color signal does not include the:

(a) chrominance and luminance signals.
(b) aft correction signal.
(c) burst signal.
(d) synchronizing and blanking signals.

11. Which of the following is not attenuated in the bandpass amplifier?

(a) The 4.5-MHz signal.
(b) The luminance signal.
(c) The color burst and synchronizing signals.
(d) The chrominance portion of the composite color signal.

12. The composite color input signal to the chroma bandpass amplifier is most likely to come from:

(a) a video amplifier.
(b) a receiver i-f amplifier.
(c) the receiver synchronizing section.
(d) the tuner.

13. A keying pulse is delivered to the chroma bandpass amplifier circuit during horizontal retrace. This keying pulse:

(a) increases the average dc level of the signal.
(b) removes the sound i-f signal.
(c) removes the luminance signal.
(d) removes the color burst.

14. The color-sync section of a color receiver:

(a) generates the 3.58-MHz reference signal and regulates its frequency and phase.
(b) turns the chroma bandpass amplifier off during trace time.
(c) keeps the three guns of the color picture tube in step.
(d) eliminates the need for a color killer circuit.

15. Which of the following statements is true regarding the I and Q signals?

(a) The frequency of the I signal is higher than the frequency of the Q signal.
(b) The I and Q signals have the same frequency, and they are 90° out of phase.
(c) The I and Q signals have the same frequency, and they are 180° out of phase.
(d) The frequency of the I signal is lower than the frequency of the Q signal.

16. Which of the following statements is true regarding the R − Y and B − Y signals?

(a) The frequency of the R − Y signal is higher than the frequency of the B − Y signal.
(b) The R − Y and B − Y signals have the same frequency, and they are 90° out of phase.

(c) The R − Y and B − Y signals have the same frequency, and they are 180° out of phase.
(d) The frequency of the R − Y signal is lower than the frequency of the B − Y signal.

17. Another way of saying that two signals are in quadrature is to say that they are:

(a) in phase.
(b) 90° out of phase.
(c) 180° out of phase.
(d) equal in frequency.

18. Two types of modulation present in the chrominance signal are:

(a) pulse position and amplitude.
(b) frequency and phase.
(c) phase and amplitude.
(d) pulse amplitude and amplitude.

19. Which of the following is correct for the luminance signal as transmitted?

(a) $E_Y = 0.30E_R + 0.59E_G + 0.11E_B$.
(b) $E_Y = 0.30E_G + 0.59E_R + 0.11E_B$.
(c) $E_Y = 0.30E_B + 0.59E_R + 0.11E_G$.
(d) $E_Y = 0.30E_R + 0.59E_B + 0.11E_G$.

20. To reproduce the color picture accurately, the color difference signals must be added to the:

(a) burst signal.
(b) 3.58-MHz oscillator signal.
(c) luminance signal.
(d) blanking pedestals.

21. In a certain receiver the chrominance signal is amplified to the desired level before it is applied to the demodulators. This is known as:

(a) quadrature-phase detection.
(b) high-level demodulation.
(c) color-amplitude demodulation.
(d) color-difference demodulation.

22. The background illumination of the three color signals is represented by:

(a) the peak voltage of the ac color signals delivered to the color CRT.
(b) the average value of the ac color signals delivered to the color CRT.
(c) the amplitude of the blanking pedestal.
(d) the amplitude of the color burst.

23. The automatic degaussing circuit of a color receiver employs a VDR and:

(a) an LDR.
(b) a varactor.
(c) a thermistor.
(d) an ABL.

24. The color *yellow* in color television is obtained by combining:

(a) green and blue.
(b) green and red.
(c) blue and red.
(d) green and white.

25. You can obtain any color with the proper amounts of hue, saturation, and:

(a) frequency.
(b) primary colors.
(c) phase.
(d) brightness.

26. In a color receiver, the saturation is determined by the:

(a) phase of the I and Q signals.
(b) amplitude of the color signal.

(c) amplitude of the luminance signal.
(d) frequency of the burst signal.

27. In Fig. 8-9, variable resistor R is a:

(a) tint control.
(b) contrast control.
(c) beam landing control.
(d) color control.

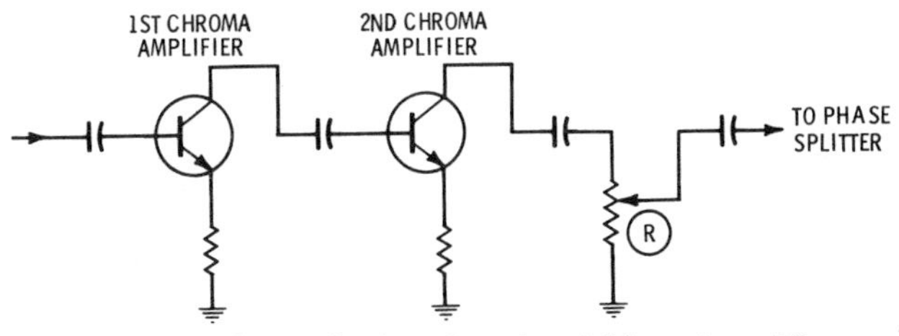

Fig. 8-9. What is the function of variable resistor R?

28. The human eye is most responsive to:

(a) purple.
(b) blue-green.
(c) green-yellow.
(d) red.

29. Transmitted colors are changed by varying the relationship between:

(a) I signals and Y signals.
(b) Q signals and E signals.
(c) I, Q, and Y signals.
(d) E_B and E_G signals.

30. In the block diagram of Fig. 8-10, the bandpass amplifier is:

(a) block "A."
(b) block "B."
(c) block "C."
(d) block "D."

31. In the block diagram of Fig. 8-10, the 3.58-MHz oscillator is:

(a) block "A."
(b) block "B."
(c) block "C."
(d) block "D."

32. In the block diagram of Fig. 8-10, the color killer is:

(a) block "A."
(b) block "B."
(c) block "C."

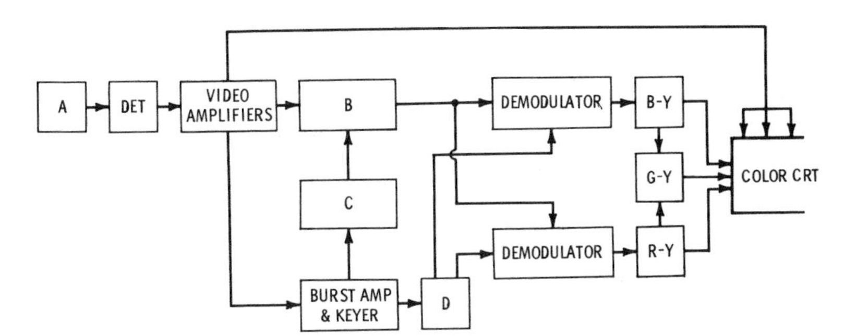

Fig. 8-10. Block diagram of the color section of a TV receiver.

33. In the frequency distribution curve of Fig. 8-11, the color subcarrier is located at:

(a) point "C."
(b) point "D."
(c) point "G."

34. In the frequency distribution curve of Fig. 8-11, the frequency range marked "E" is:

(a) 0.75 MHz.
(b) 3.58 MHz.
(c) 4.2 MHz.
(d) 4.5 MHz.

35. In the frequency distribution curve of Fig. 8-11, the point marked "H" should (ideally) be:

(a) 4.2 MHz above "D."
(b) 3.58 MHz above "G."
(c) 2.75 MHz above "C."

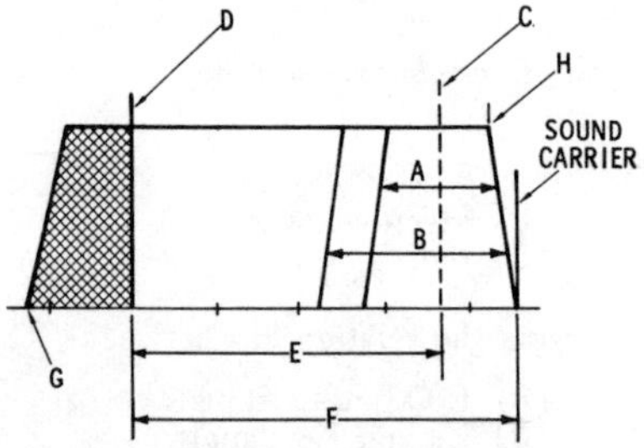

Fig. 8-11. Frenqucy distribution of a television signal.

36. In the frequency distribution curve of Fig. 8-11:

(a) the I signal is represented by "A."
(b) the I signal is represented by "B."

37. In the frequency distribution curve of Fig. 8-11, the frequency value 4.5 MHz is associated with:

(a) range "F."
(b) point "H."
(c) range "E."
(d) range "B."

38. The block diagram of Fig. 8-12 shows:

(a) an agc system.
(b) an aft system.
(c) an acc system.
(d) an abl system.

39. In the block diagram of Fig. 8-12, which of the following could be substituted for the varactor diode?

(a) A varistor.
(b) A zener diode.
(c) A reactance tube.
(d) A slope detector.

40. Fig. 8-13 shows the overall rf and i-f response of a color receiver. The video carrier should be located at:

(a) point "A."
(b) point "B."
(c) point "C."
(d) point "D."

41. In the overall response curve of Fig. 8-13, the sound carrier should be located at:

(a) point "A."
(b) point "B."
(c) point "C."
(d) point "D."

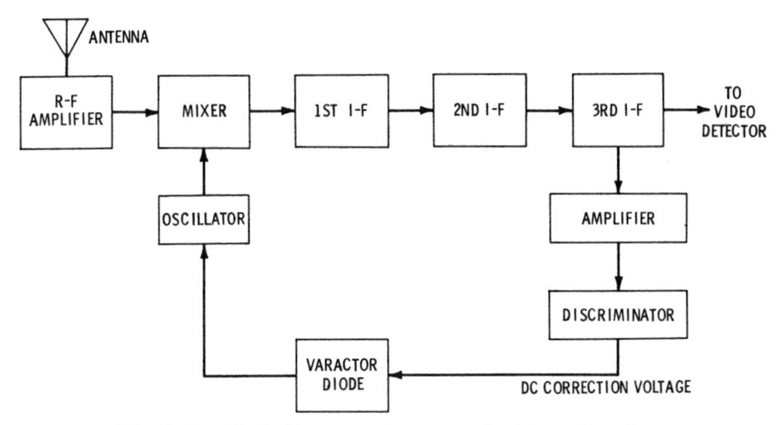

Fig. 8-12. Block diagram of tuner and video i-f section.

42. The gradual slope from point "C" to point "E" on the curve of Fig. 8-13 is necessary to compensate for the:

(a) vestigial sideband.
(b) additional signal strength added by the color carrier.
(c) poor upper-frequency response of transistors.
(d) excessive high-frequency response of video i-f transformer coupling.

43. In the circuit of Fig. 8-14:

(a) R1 is the tint control.
(b) R2 is the tint control.
(c) the tint control is not shown.

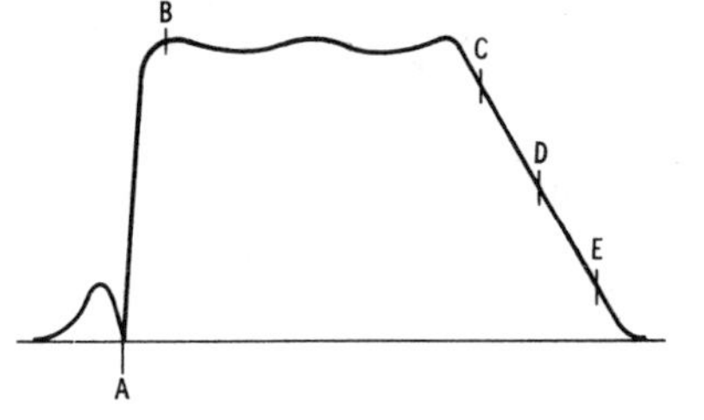

Fig. 8-13. Overall rf and i-f response curve.

44. The lateral-correction magnet moves:

(a) the red beam.
(b) the blue beam.
(c) the green beam.

45. The high-frequency response of the video amplifier is set by the:

(a) contrast control.
(b) brightness control.
(c) video peaking contol.
(d) CRT bias control.

46. In a certain color receiver the video amplifiers are direct-coupled to the CRT. Which of the following could not be located in the luminance channel?

(a) The brightness control.
(b) The contrast control.
(c) The video peaking control.
(d) The agc bias control.

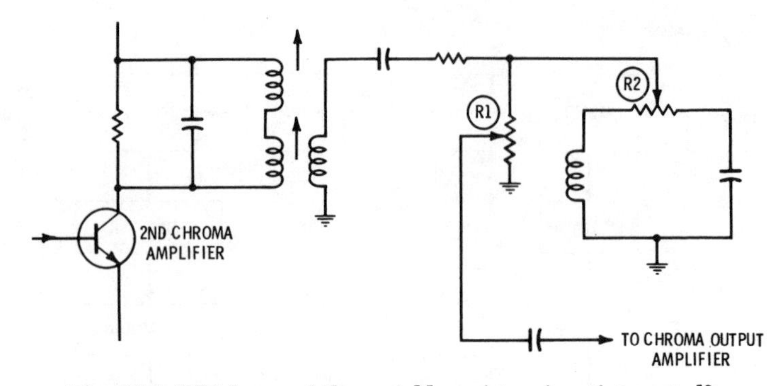

Fig. 8-14. Which one of the variable resistors is a tint control?

47. A process by which two signals can modulate two carriers having the same frequency, but having phase angles which differ by 90° is called:

(a) double-sideband modulation.
(b) divided-carrier modulation.
(c) phase modulation.
(d) frequency modulation.

48. When three beams of a color picture tube converge at the aperture mask as they are deflected vertically and horizontally, the process is called:

(a) static convergence.
(b) aperture laying.
(c) dynamic convergence.
(d) bias control.

49. Illuminant C is:

(a) the standard of measurement for red color.
(b) a standard for green-yellow colors.
(c) a type of phosphor used in color picture tubes.
(d) the reference white of color television.

50. A type of demodulator that utilizes a reference signal which has the same frequency as the carrier (or subcarrier) of the signal being demodulated is:

(a) a reactance demodulator.
(b) a synchronous demodulator.
(c) an orthographic demodulator.
(d) a slope detector.

9

The Synchronizing Circuits

KEYED STUDY ASSIGNMENTS

Howard W. Sams Photofact Television Course
Chapter 7—Sawtooth Generator Control and Production of Scanning Waveforms
Chapter 8—Deflection Systems
Chapter 10—Sync-Pulse Separation, Amplification, and Use
Howard W. Sams Color-TV Training Manual
Chapter 5—Monochrome Circuitry

This assignment covers sweep circuits that are used in both monochrome and color receivers.

IMPORTANT CONCEPTS

Fig. 9-1 shows a block diagram of the vertical and horizontal sweep sections in a television receiver. The sync take-off point in this illustration is at the first video amplifier, but in some receivers it may be in a later video-amplifier stage where more signal amplitude is available.

A noise inverter is shown at the sync take-off point. Its purpose is to eliminate noise that could interfere with sweep-oscillator synchronization. This is especially important in the horizontal oscillator section because short-duration noise pulses could trigger the horizontal oscillator before the horizontal pulse arrives from the differentiator.

Fig. 9-2 shows the principle of operation of the noise-inverter section. The composite video signal from the detector contains noise

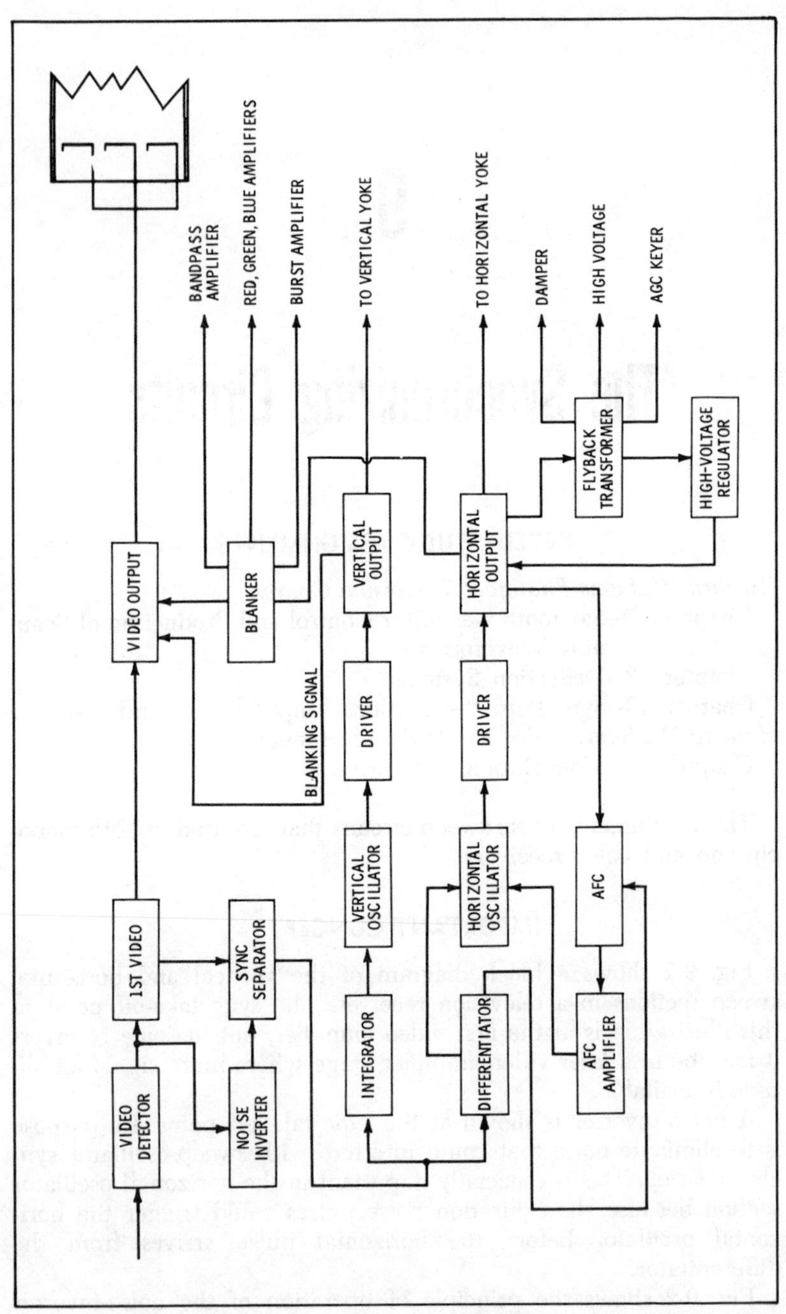

Fig. 9-1. Block diagram of the sweep system of a TV receiver.

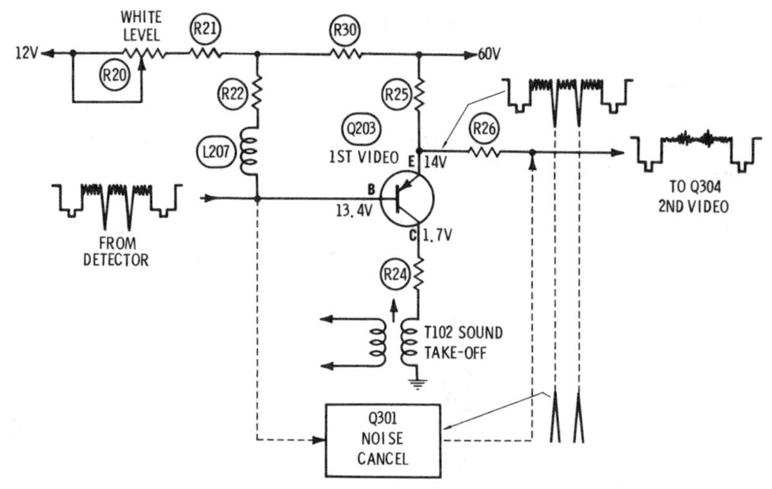

Fig. 9-2. Noise inverter reverses phase of noise spikes.

spikes which are delivered to the first video amplifier. At the same time, this signal is delivered to a noise-cancelling circuit which inverts the phase of the noise spikes. Note that the first video amplifier is an emitter follower, and therefore, no phase inversion of the signal takes place. At the output of the video amplifier the inverted noise spikes are recombined with the composite signal. The output is shown with the spikes removed.

Fig. 9-3 shows a typical noise-cancelling circuit. The transistor in this stage is normally cutoff because of the reverse bias on the emitter and the base. It remains cut off until a noise spike of sufficient amplitude overrides the reverse bias. This requires that the noise spikes have a greater amplitude than the sync pulse in a composite signal. The noise spikes forward bias the transistor and cause it to conduct and amplify the peaks. The bias is adjusted so that sync pulses will not trigger the stage into conduction. The diode isolates the noise-cancelling stage from the detector output circuit.

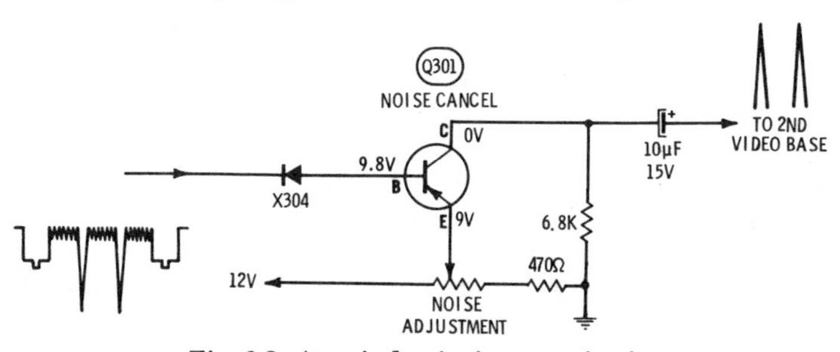

Fig. 9-3. A typical noise-inverter circuit.

Besides the problem of noise spikes, it is also possible for hum voltages to arrive with the composite video signal as it is delivered to the sync separator. This is especially true in tube-type receivers where a heater-to-cathode short may occur in the i-f stage. Fig. 9-4 shows how the raster is distorted when 60 hertz or 120 hertz hum voltages are present in the horizontal-sweep circuit.

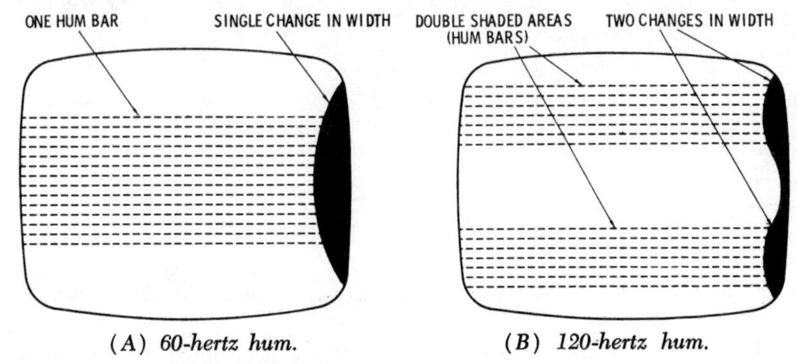

(A) 60-hertz hum. *(B) 120-hertz hum.*

Fig. 9-4. Distortion caused by hum voltage.

There are two output signals from the sync separator. They go to the integrator and differentiator of the vertical and horizontal sweep circuits, respectively. Fig. 9-5 shows a simplified drawing of the integrating and differentiating networks. Note that the integrator is a low-pass filter. Its purpose is to add (or integrate) the pulses on the vertical blanking pedestal in order to produce a trigger for the vertical deflection oscillator.

Either a blocking oscillator or a multivibrator circuit may be used for the vertical oscillator. Fig. 9-6 shows a typical circuit. In this case a blocking oscillator circuit is used. The dc bias on the oscillator stage is set by adjustment of the vertical hold control. Diodes D2 and D8 short circuit when the field in transformer T1 is collapsing. These diodes serve to dissipate quickly the energy stored in the magnetic field.

The frequency of oscillation is determined primarily by the time constant of C5, R9, and the vertical hold control. This is a free-running oscillator which is synchronized by a vertical sync pulse input to the base of the transistor. In order for synchronization to occur, the free-running frequency of the oscillator must be lower than the frequency of the sync pulses. Thus, the sync pulses trigger the transistor into conduction slightly before it would normally conduct as a free-running oscillator. The output waveform of the blocking oscillator is modified by a feedback signal from the vertical output stage delivered through the vertical linearity control.

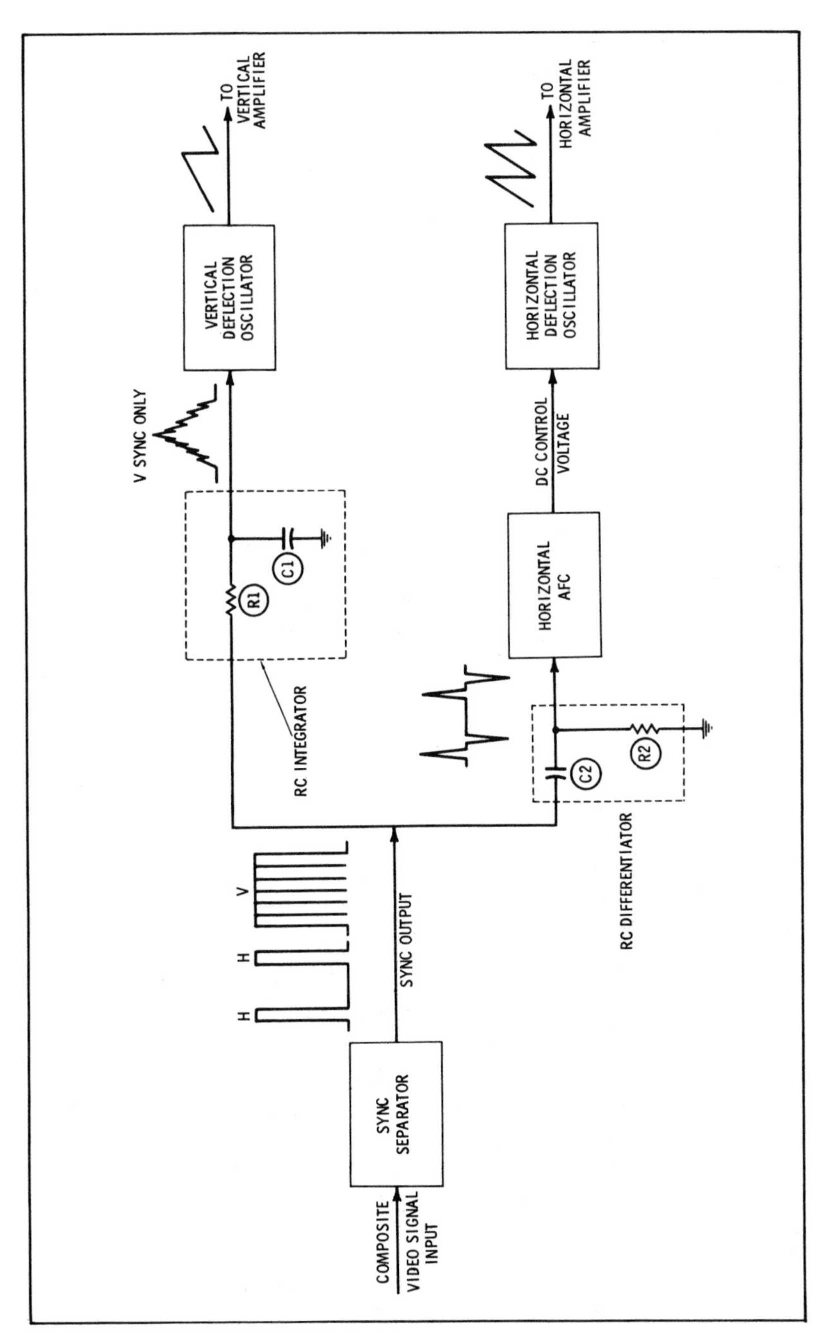

Fig. 9-5. Simplified integrator and differentiator circuits.

The driver stages following both the vertical oscillator and the horizontal oscillator are necessary in order to present a high impedance to the oscillator output. At the same time, the driver presents a low impedance to the vertical and horizontal output stages. Fig. 9-7 shows a typical vertical driver and vertical output stage. Notice that the vertical output amplifier operates directly into the yoke and an output transformer is not required. This is characteristic of transistor circuits which have a relatively low output impedance. (The same technique is sometimes used in audio stages where the output transistor delivers the audio signal directly to a speaker without the use of an output transformer.)

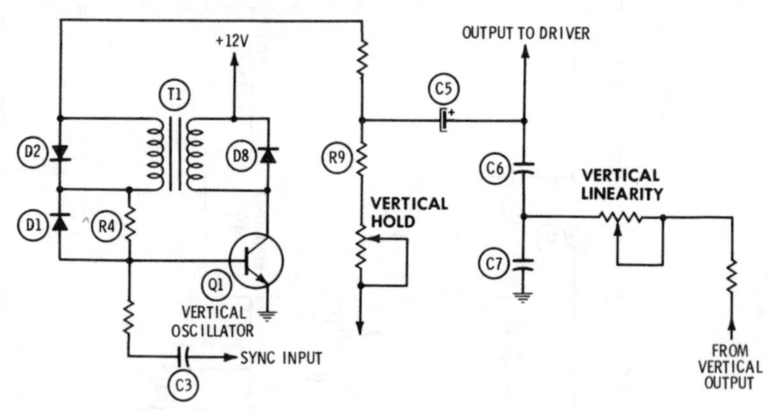

Fig. 9-6. A typical transistor vertical oscillator circuit.

Since the transistor amplifier is direct coupled to the yoke, an isolation capacitor (C10) is needed to prevent dc current in the yoke and thus decenter the picture. The capacitor shown here is typical for most transistor circuits.

The vertical output-transistor load impedance is L1, which develops the high voltage necessary for retrace. The gain of the power output stage is controlled by the height control, while the vertical linearity control regulates the amount of feedback signal used for correcting the output waveform of the vertical oscillator. The vertical output stage is stabilized from temperature changes by thermistor R7. The highly positive pulse at the collector of the vertical output stage reverse biases zener Z4 and produces a vertical blanking signal. This blanking signal is delivered to the video output stage as shown in the block diagram of Fig. 9-1. This is possible if the video output stage is direct coupled with the CRT. If not, then, the blanking signal will go to the biasing stage of the CRT.

The output of the differentiator in Fig. 9-1 goes to the afc circuit, which also has an input signal from the horizontal output stage.

The purpose of the afc circuit is to compare these two signals, and produce a dc correction voltage which controls the oscillator frequency. The afc amplifier is a dc amplifier stage which is direct coupled to the oscillator.

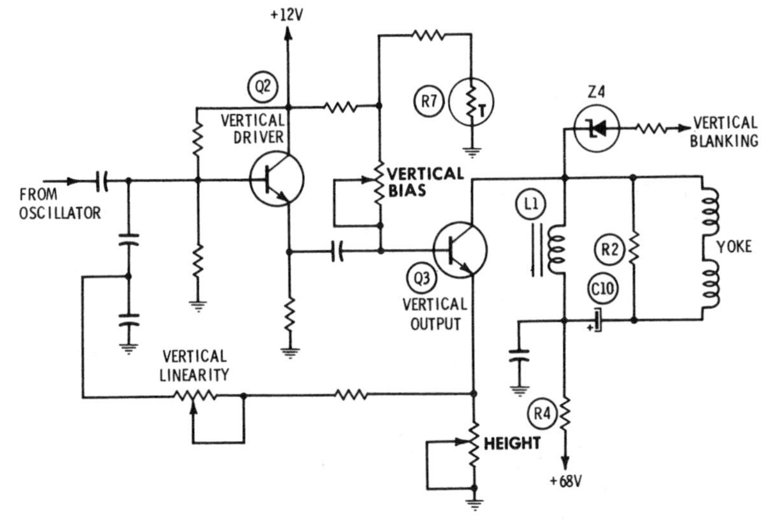

Fig. 9-7. A vertical driver and output stage.

Fig. 9-8 shows a typical afc amplifier and oscillator stage. The input signal to the afc circuit is comprised of a sync pulse from the sync separator and a signal from the flyback transformer. In other circuits, the oscillator feedback signal (to be compared with the sync pulse) comes from the horizontal output stage. If the timing between the sync pulse and the feedback signal from the horizontal output is not correct, a dc correction voltage is delivered to the afc amplifier. This is a simple dc amplifier with its collector direct coupled to the base of the oscillator. Feedback from the blocking oscillator transformer is through capacitor C24. This capacitor (or more precisely, the discharge rate of this capacitor) determines the frequency of the horizontal oscillator. Diode X7 protects the emitter-base junction of the oscillator stage from excessive reverse bias during the cutoff period.

The output from the oscillator is delivered to a driver stage which is transformer coupled to the horizontal output stage. The duration of the pulse from the oscillator is only about 25 microseconds, which represents the time for about one-half of the horizontal scan. A sawtooth waveform appears at the base of the output transistor because of capacitor C27. If it were not for this capacitor, the waveform at the base would be nearly a square wave.

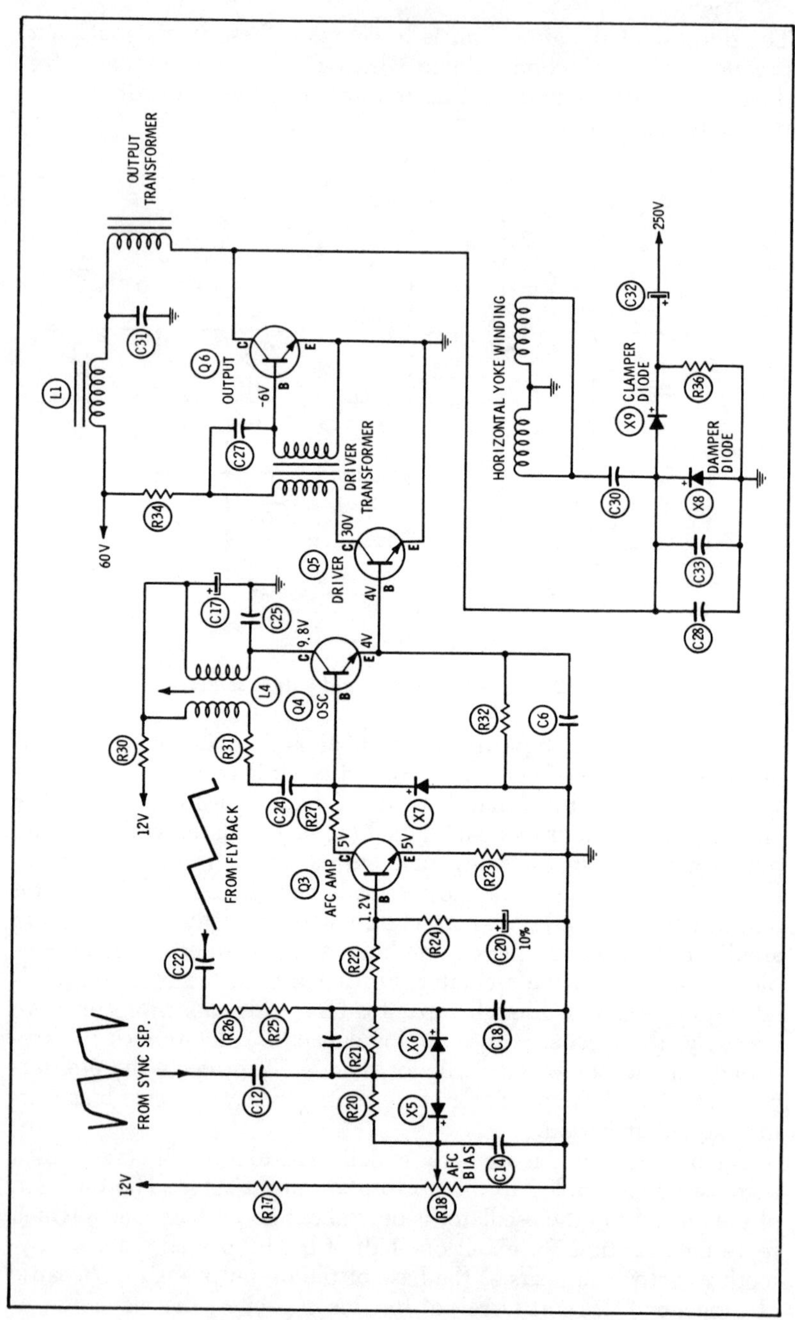

Fig. 9-8. A typical horizontal sweep circuit.

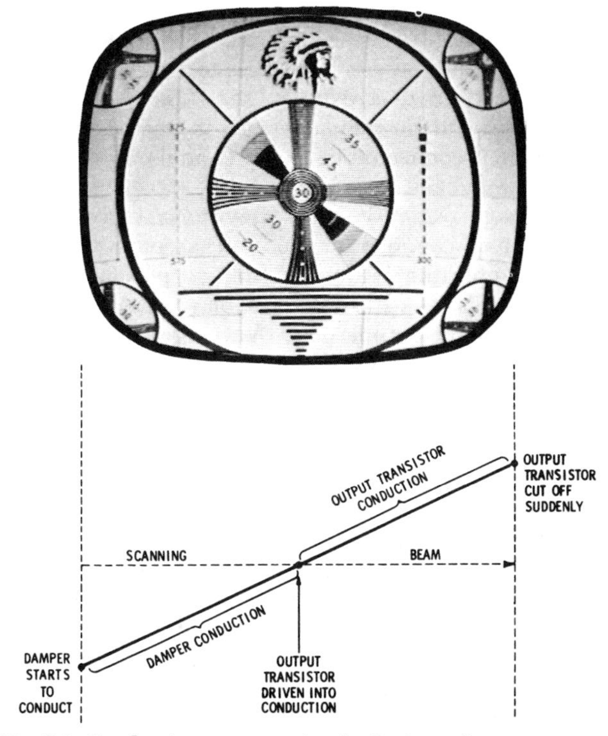

Fig. 9-9. Conduction sequence for the horizontal output circuit.

The horizontal output stage does not produce the signal for the full horizontal sweep. One-half of the sweep is produced by output transistor conduction, and the other half is produced during damper conduction. Fig. 9-9 shows how the horizontal sweep is divided. Notice that there is a point at the center of the sweep when damper conduction ceases and output transistor conduction begins. Adjustment of the horizontal drive control normally sets the proper transition from one type of conduction to the other. If the drive control is improperly set, a vertical white line will appear in the center of the picture, representing an overlap of the sweep signals.

The output transistor delivers its signal to a damper diode X8, which is a semiconductor diode in this case. Note again that an output transformer is not used, and therefore, capacitor C30 is used to prevent dc coupling into the horizontal yoke winding. The clamper diode X9 protects the output transistor in the case of a momentary arc in the high-voltage section.

In a variation of the horizontal circuitry, a varactor diode is used in place of the afc amplifier for controlling the horizontal oscillator frequency.

Returning again to the block diagram of Fig. 9-1, the output of the horizontal output stage is delivered to the flyback transformer which has connections to a damper, high-voltage rectifier, an agc keyer, and a high-voltage regulator. (In the circuit just described, the damper stage is connected to the horizontal output section rather than through the flyback transformer.)

The high-voltage section in color television receivers is regulated, and there are two common methods of accomplishing this. One method is to sample the amount of voltage present in the flyback transformer, rectify it, and use it for a bias voltage to regulate the gain of the horizontal output stage. Varying the amplitude of the signal from the horizontal output stage will also vary the amount of high voltage. Thus, the feedback method of high-voltage regulation consists of a negative, closed-loop feedback circuit which is similar to agc and avc systems.

The other method of regulating the high voltage is to use a shunt regulator across the high-voltage line.

The composite signal shown in Fig. 9-10 illustrates that the burst signal goes well into the video range, below the black level, during portions of each cycle. This means that during the horizontal retrace the burst signal will have a sufficient amplitude to drive the CRT and produce illumination on the screen. The indication is usually a narrow yellow region on the right side of the color CRT screen. For this reason, retrace on the horizontal line is blanked in addition to blanking on the vertical retrace. Fig. 9-1 shows that the input to the blanker stage comes from the horizontal output stage, but in some receivers it comes from the flyback transformer. The blanker delivers the cutoff signal to the video amplifier, which biases the

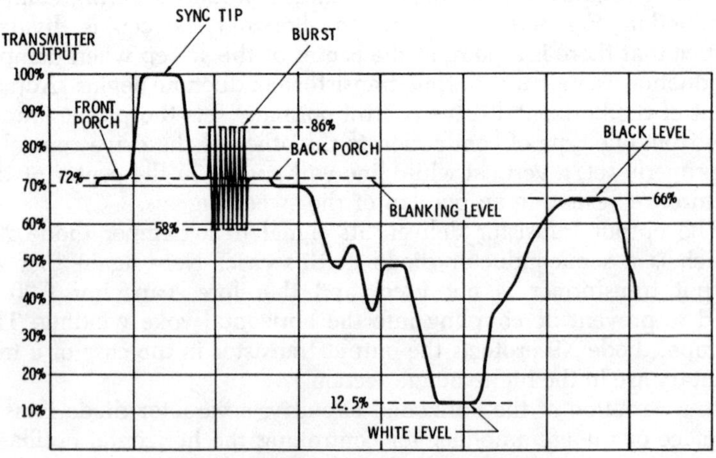

Fig. 9-10. Relative position of burst signal.

CRT to cutoff during horizontal retrace. As in the case of the vertical blanking signal, it is necessary that the video stage be direct coupled to the CRT in order for the blanking signal to enter at the video amplifier stage. Otherwise, the blanking signal will enter at the bias circuitry of the CRT.

The blanker stage also delivers a signal to the bandpass amplifier to key that amplifier off during periods when the burst is present. In the block diagram of Fig. 9-1 there is an output from the blanker to the red, green, and blue amplifiers. This particular output is used for dc restoration, but it is not present in all color receivers. Another output signal from the blanker goes to the burst amplifier to key that amplifier on during the presence of the burst. The burst amplifier is cut off at all other times.

PROGRAMMED QUESTIONS AND ANSWERS

Starting with question number 1, select the answer that you feel is correct. If you feel that (A) is correct, proceed to block number 17 as directed. If you feel that (B) is correct, proceed to block number 9 as directed. If you feel that more than one answer is correct, choose the one that you think is the *most* correct.

. . . .

1 Distortion in the raster, as shown in Fig. 9-11, is normally caused by:

(A) a 60-Hz ripple in the horizontal sweep circuit. (Go to block number 17.)

(B) a 120-Hz ripple in the horizontal sweep circuit. (Go to block number 9.)

. . . .

2 Your answer is wrong. Review Fig. 9-9, then go to block number 10.

. . . .

3 The integrator circuit is a low-pass filter. The capacitor would short high-frequency signals to ground. The RC integrator

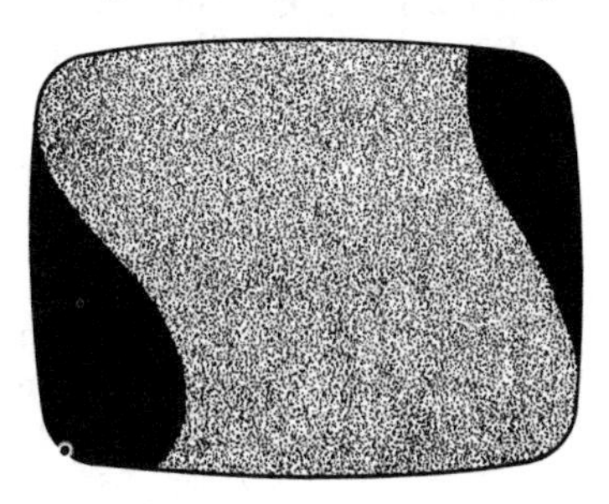

Fig. 9-11. Raster distortion.

circuit is not designed for that purpose, but its configuration is that of a low-pass filter. High-frequency noise pulses will be grounded in the integrator, and this is one of the reasons why the horizontal sweep circuit is more affected by noise spikes than the vertical sweep circuit.

Here is the next question . . .

A variable resistor is used to control the gain of the vertical output stage. Adjustment of this resistor will:

(A) stop the picture from rolling up or rolling down. (Go to block number 24.)

(B) change the height. (Go to block number 16.)

. . . .

4 Your answer is wrong. The flyback transformer does not determine the waveshape of the signal to the yoke. Therefore, a change in the shape of the raster cannot be caused by the flyback transformer Go to block number 14.

. . . .

5 An emitter follower, such as the type used for a driver, has a high input impedance and a low output impedance.

Here is the next question . . .

A thermistor is used in the yoke to:

(A) protect it from high-voltage kickback. (Go to block number 12.)

(B) compensate for temperature changes. (Go to block number 22.)

. . . .

6 This circuit is used for a sync separator. The transistor emitter-base junction is reverse biased, so the transistor is cut off. Only the positive sync pulses have a sufficient amplitude to forward bias the emitter-base junction and produce an output pulse.

Here is the next question . . .

The integrator is:

(A) a high-pass filter. (Go to block number 11.)

(B) a low-pass filter. (Go to block number 3.)

. . . .

7 The vertical white line is produced when the yoke is defective.

Here is the next question . . .

To assure that the burst amplifier conducts only when there is a burst signal present, a gate signal is sent from the__________.
(Go to block number 26.)

. . . .

8 Your answer is wrong. If the horizontal output stage is defective, there would be no high voltage and you could not see the vertical white line. Go to block number 7.

. . . .

9 Your answer is wrong. A 120-Hz ripple would show as two complete cycles on the right side of the raster. Go to block number 17.

. . . .

10 As shown in Fig. 9-9, the output amplifier conducts during the second half of the line.

Here is the next question . . .

If the horizontal retrace is not blanked:

(A) the burst will cause a yellow stripe on the right side of the picture. (Go to block number 23.)

(B) there will be no symptom since the burst is on top of the blanking pedestal. (Go to block number 15.)

. . . .

11 Your answer is wrong. Go to block number 3.

. . . .

12 Your answer is wrong. A thermistor is a temperature-sensitive resistor. It is not used for high-voltage protection. Go to block number 22.

. . . .

13 Although the input to the afc circuit does not come directly from the horizontal oscillator, it does compare the sync pulse with the oscillator frequency. Usually the oscillator signal is amplified before it is delivered to the afc circuit.

Here is the next question . . .

A driver stage often follows the vertical and horizontal oscillators. The driver presents a:

(A) high impedance to the oscillator circuit. (Go to block number 5.)

(B) low impedance to the oscillator circuit. (Go to block number 20.)

. . . .

14 A defective yoke will cause a keystone raster.

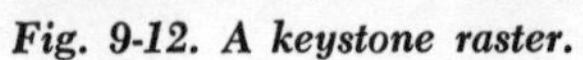

Fig. 9-12. A keystone raster.

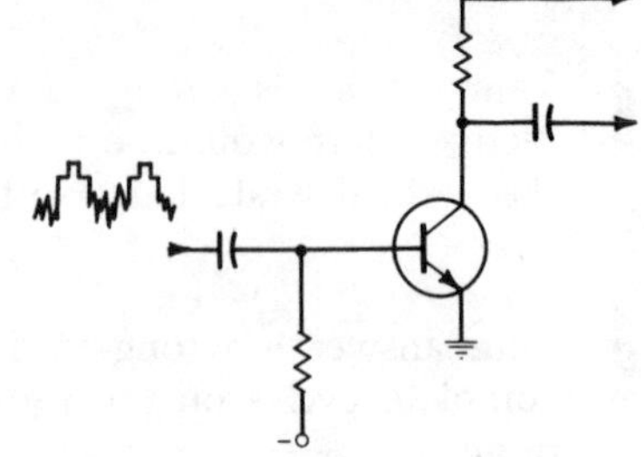

Fig. 9-13. A transistor amplifier.

Here is the next question . . .
The transistor circuit of Fig. 9-13 is:
(A) biased beyond cutoff. (Go to block number 6.)
(B) conducting to saturation. (Go to block number 19.)

15 Your answer is wrong. The burst extends into the video portion of the signal shown in Fig. 9-10. Go to block number 23.

16 Changing the gain of the vertical output stage will change the height of the picture.
Here is the next question . . .
A capacitor is often placed in series with the yoke winding in transistor vertical output or horizontal output stages. The purpose of this capacitor is to:
(A) isolate the yoke from the dc voltages in the amplifier. (Go to block number 18.)
(B) compensate for the low impedance of a transistor amplifier. (Go to block 25.)

17 The curve shows one complete cycle of ripple on the raster and this represents a ripple frequency of 60 Hz.
Here is the next question . . .
The afc circuit compares a pulse signal from the differentiator and:
(A) the integrator. (Go to block number 21.)
(B) the horizontal oscillator. (Go to block number 13.)

18 It is important to keep the dc out of the yoke, since this will affect the centering of the raster.
Here is the next question . . .

The horizontal output stage conducts:
(A) during the first half of the line. (Go to block number 2.)
(B) during the second half of the line. (Go to block number 10.)

. . . .

19 Your answer is wrong. A positive voltage is needed on an npn transistor to bias it into conduction. Go to block number 6.

. . . .

20 Your answer is wrong. It is desirable to present a high impedance to the oscillator in order to keep from loading it and affecting its output frequency. Go to block number 5.

. . . .

21 Your answer is wrong. The integrator adds the pulses on the vertical blanking pedestal and triggers the vertical oscillator. Go to block number 13.

. . . .

22 Changes in yoke temperature cause changes in the winding resistance. In turn, this causes changes in the current waveshape when a trapezoidal voltage is generated across the coil. The thermistor changes its resistance when the temperature rises, and the change compensates for the change in yoke resistance.
Here is the next question . . .
A keystone raster, as shown in Fig. 9-12, is caused by a defective:
(A) flyback transformer. (Go to block number 4.)
(B) yoke. (Go to block number 14.)

. . . .

23 The blanker circuit delivers a cutoff signal to the video amplifier (or CRT bias circuit) to prevent the burst from reaching the screen during retrace.
Here is the next question . . .
A vertical white line in the center of the screen, and no raster, indicates that:
(A) the yoke is defective. (Go to block number 7.)
(B) the horizontal output stage is not amplifying. (Go to block number 8.)

. . . .

24 Your answer is wrong. The vertical hold control sets the oscillator frequency. This control is in the oscillator circuit, not in the vertical output stage. Go to block number 16.

. . . .

25 Your answer is wrong. The low impedance of the transistor amplifier is an advantage for coupling a signal to the yoke. Go to block number 18.

. . . .

26 To assure that the burst amplifier conducts only when there is a burst signal present, a gate signal is sent from the *blanker, or from the horizontal output circuit.*

You have now completed the programmed questions and answers.

. . . .

PRACTICE TEST

1. The free running frequency of a sweep oscillator should be:

(a) slightly above the synchronized frequency.
(b) slightly below the synchronized frequency.
(c) exactly equal to the synchronized frequency.

2. Which of the following is least likely to be used as a sweep oscillator in a television receiver?

(a) An UJT oscillator.
(b) A neon sawtooth oscillator.
(c) A transistor blocking oscillator.
(d) A transistor multivibrator.

3. To prevent ringing and excessively high voltages from being generated by the collapsing magnetic field in the flyback transformer:

(a) the output winding is made with many more turns than necessary.
(b) a neon lamp is placed across the winding.
(c) a damper is used.
(d) a blocking oscillator, rather than a multivibrator, is used for the sweep oscillator.

4. Which of the following statements is correct?

(a) There are two fields per frame.
(b) There are two frames per field.

5. The total number of horizontal lines transmitted per second is:

(a) 30.
(b) 60.
(c) 525.
(d) 15,750.

6. A keystone raster is caused by a:

(a) 60-hertz ripple in the horizontal sweep.
(b) 120-hertz ripple in the horizontal sweep.
(c) defective yoke.
(d) shorted burst amplifier input circuit.

7. **The keying pulse for the burst amplifier may come directly from the flyback transformer, or from the:**

   (a) blanker.
   (b) sync separator.
   (c) horizontal oscillator.
   (d) yoke.

8. **A noise inverter is usually used:**

   (a) to keep snow out of the picture.
   (b) to prevent noise spikes from reaching the horizontal oscillator.
   (c) for testing the fm discriminator.
   (d) to blank the vertical retrace on the CRT.

9. **To determine if the high-voltage circuit is working, you should:**

   (a) rock the horizontal output tube in its socket and see if white flashes appear on the screen.
   (b) see if the high-voltage lead will arc to the chassis.
   (c) place a temporary short across the damper.
   (d) measure the boost voltage.

10. **The input to the blanker circuit comes from:**

    (a) the video detector.
    (b) one of the video amplifiers.
    (c) the horizontal output or flyback.
    (d) the sync separator.

11. **If the horizontal retrace is not blanked, the burst may show on the screen as:**

    (a) colored speckles of snow.
    (b) a color stripe on the right side of the screen.
    (c) horizontal green stripes in the picture.
    (d) excessive contrast.

12. **Which of the following input or output signals would not normally be expected at the sync separator?**

    (a) An input signal from video detector.
    (b) A dc bias voltage from the agc circuit.
    (c) An output signal to a horizontal differentiator.
    (d) An output signal to a vertical differentiator.

13. **Which of the following is a primary purpose of the *vertical switch transistor* in the vertical sweep circuit?**

    (a) To energize the vertical sweep circuit when the receiver is first turned on.
    (b) To start the vertical sweep oscillator when the receiver is first turned on.
    (c) To discharge the integrator capacitor.
    (d) To turn the vertical sweep off when high-amplitude noise spikes are present.

14. **For top and bottom pincushion correction:**

    (a) a signal from the horizontal yoke circuit is coupled to the vertical yoke circuit.
    (b) a signal from the vertical yoke circuit is coupled to the horizontal yoke circuit.
    (c) the B+ supply voltage is varied in the horizontal oscillator circuit.
    (d) the B+ supply voltage is varied in the vertical oscillator circuit.

15. A blocking oscillator is most nearly like:

(a) a Clapp oscillator.
(b) a Colpitts oscillator.
(c) an Armstrong oscillator.
(d) a multivibrator.

16. A multivibrator circuit cannot be synchronized by:

(a) a sine-wave voltage on the tube grid or transistor base.
(b) a pulse on the tube grid or transistor base.
(c) the boost from the flyback on the tube grid or transistor base.

17. Feedback in a blocking oscillator is through a:

(a) coupling capacitor.
(b) saturated transistor or tube.
(c) transformer.
(d) diode.

18. A square-wave voltage across a pure inductance will cause a:

(a) sawtooth (or triangular) current through the coil.
(b) square-wave current through the coil.
(c) sinusoidal current through the coil.
(d) parabolic current through the coil.

19. Which of the following pulses controls interlace?

(a) Equalizing pulses.
(b) Horizontal pulses from the flyback.
(c) Horizontal afc circuit.
(d) Burst.

20. The right half of each horizontal line is the result of:

(a) current from the horizontal output stage.
(b) decay current through the damper.
(c) an amplified sync pulse.
(d) a signal from the blanker.

21. The width of the raster is controlled:

(a) by varying the filament current in the horizontal output stage.
(b) by varying the height of the sync pulse from the sync separator.
(c) by means of a variable inductor which shunts a portion of the secondary of the horizontal output transformer.
(d) by varying the boost voltage.

22. When the base of a transistor is at the same voltage as the emitter, the transistor is:

(a) saturated.
(b) operating normally.
(c) cut off.

23. Meter damage may result from measuring the:

(a) boost voltage.
(b) plate voltage on a horizontal or vertical output tube.
(c) collector voltage of a sync separator.
(d) emitter-to-base voltage on a transistor multivibrator.

24. The output of the afc phase diodes:

(a) consists of pulses for controlling the frequency of the horizontal sweep oscillator.
(b) is a dc voltage when there is a difference in frequency between the sync pulse and the waveform from the flyback.

(c) is never zero volts dc.
(d) is a pure sine-wave voltage.

25. In a certain tube-type receiver, the boost voltage measures below normal, but it rises to normal when the yoke is disconnected. Which of the following is correct?

(a) This is an indication that the oscillator frequency is too low.
(b) This is an indication that there is a heater-to-cathode short in the horizontal output tube.
(c) This is an indication that the low-voltage supply has a defective filter.
(d) This is an indication that the yoke is defective.

26. The most likely cause of a keystone raster is:

(a) poor low-voltage filtering.
(b) insufficient drive.
(c) excessive drive.
(d) a defective yoke.

27. When the vertical oscillator frequency is too high:

(a) the picture will roll up.
(b) the picture will roll down.
(c) a pincushion raster will result.
(d) the sync separator is defective.

28. A certain vacuum-tube triode is being used as a switch to discharge a capacitor. The plate of the triode is connected to one capacitor plate, and the cathode is connected to the other capacitor plate. The switch is closed when the:

(a) grid is highly negative with respect to the cathode.
(b) tube is saturated.
(c) cathode is positive with respect to ground.
(d) minimum amount of grid current flows.

29. In a tube-type receiver, disconnecting the plate cap of the horizontal output stage will:

(a) isolate a trouble as being either before or after the horizontal output stage.
(b) cause the boost voltage to rise if the stage is operating properly.
(c) cause excessive screen current to flow in the tube.
(d) show that the stage is defective if the fuse blows.

30. You can buy an adapter for adding capacitance to the damper tube. This added capacitance is sometimes needed for:

(a) decreasing the boost voltage to its normal value.
(b) rejuvenating an aging picture tube.
(c) correcting width or horizontal linearity problems.
(d) improving horizontal stability.

31. A yoke winding will normally change resistance with changes in temperature. To compensate for this:

(a) a thermistor is placed across the input to the oscillator circuit.
(b) a thermistor is placed across the yoke winding.
(c) the yoke is made with the ends of the winding flared outward.
(d) part of the yoke is made with copper wire and the rest is made with aluminum wire.

32. An integrator is a:

(a) low-pass filter.
(b) high-pass filter.
(c) circuit that combines two different kinds of oscillator operation.
(d) a circuit that combines two different kinds of pulses.

33. An open heater in the damper tube will cause:

(a) foldover on the right side of the screen.
(b) foldover at the top of the screen.
(c) loss of brightness.
(d) loss of sync.

34. The top of the picture is stretched out. You should first:

(a) adjust the vertical linearity control.
(b) replace the vertical driver transistor.
(c) replace the blocking oscillator transformer.
(d) replace the yoke.

35. Which of the stages does not directly contribute to horizontal deflection?

(a) Afc.
(b) Horizontal output stage.
(c) Horizontal oscillator.
(d) Yoke.

36. Which of the following is not likely to happen if the horizontal oscillator circuit stops producing an output pulse?

(a) A vertical white line will appear in the center of the screen.
(b) There will be a loss of brightness.

37. Loss of the horizontal sync pulse from the sync separator will:

(a) cause the picture to move out of horizontal lock.
(b) cause the oscillator to stop running.
(c) cause a vertical white line to appear in the center of the screen.

38. Barkhausen lines are:

(a) caused by oscillation in the sync separator stage.
(b) not a problem in transistor receivers.
(c) caused by mechanical vibration of the yoke.
(d) produced when there is insufficient horizontal drive.

39. A certain receiver will not hold sync horizontally or vertically. Other than this, the picture looks normal. A likely cause is:

(a) an improperly adjusted fine tuning control.
(b) a defective yoke.
(c) a defective sync separator.
(d) the diode detector was installed backwards.

40. The rectangular pattern of light on the screen is known as the:

(a) brightness.
(b) raster.
(c) screen.
(d) white square.

41. Two types of high-voltage regulators are in use. One is the shunt regulator across the high-voltage output, and the other is a closed-loop feedback system that controls the:

(a) dc bias on the oscillator.
(b) gain of the horizontal output stage.

(c) gain of the afc amplifier.
(d) gain of the blanker.

42. An example of a relaxation oscillator is a:

(a) multivibrator.
(b) synchratron.
(c) Clapp oscillator.
(d) crystal-controlled Colpitts oscillator.

43. In order to get a sawtooth current flowing through a magnetic deflection coil, the voltage across the coil will be a:

(a) trapazoidal waveform.
(b) sawtooth waveform.
(c) triangular waveform.
(d) parabolic waveform.

44. Snivets are caused by:

(a) a problem in the damper stage.
(b) oscillation in the horizontal output stage.
(c) power supply ripple in the oscillator stage.
(d) none of these.

45. A horizontal white line across the screen can be caused by:

(a) a shorted sync separator.
(b) an open integrator.
(c) a defective vertical oscillator.
(d) none of these.

46. In a color receiver, the picture on the screen is tilted. Which of these statements is true?

(a) It can be straightened by turning the picture tube.
(b) This indicates that the picture tube is defective.
(c) The yoke balance control is out of adjustment.
(d) The yoke should be tilted.

47. The burst amplifier is keyed into conduction during:

(a) the time when equalizing pulses are present.
(b) the time when the VITS is present.
(c) the time the back porch of the horizontal blanking pedestal is present.
(d) monochrome signal reception only.

48. A keying pulse may be delivered to a keyed agc circuit from the:

(a) vertical output stage.
(b) sync separator.
(c) detector.
(d) flyback transformer.

49. The sync pulse on the composite video signal is:

(a) 25% of the maximum signal amplitude.
(b) 25% of the video signal amplitude.
(c) 75% of the maximum signal amplitude.
(d) 15% of the maximum signal amplitude.

50. A white vertical line in the center of an otherwise normal picture indicates that you should:

(a) decrease the drive voltage.
(b) replace the yoke.
(c) reduce the gain of the driver.
(d) replace the oscillator tube or transistor.

10

Troubleshooting Techniques

KEYED STUDY ASSIGNMENTS

Howard W. Sams Color-TV Training Manual
Chapter 10—Setup Procedure
Chapter 11—Aligning the Color Receiver
Chapter 12—Troubleshooting the Color Receiver

The keyed study assignments are related to troubleshooting in color television receivers. This chapter concentrates on important concepts related to troubleshooting in transistorized equipment.

IMPORTANT CONCEPTS

Out-of-Circuit Tests for Transistors

In the normal operation of a transistor, the collector-base junction is reverse biased. As such, it behaves very much like a reverse-biased diode. When an inverse voltage is placed across the junction, there is small amount of reverse current flow. This reverse current flow is an indication of the condition of the junction. A large reverse current flow indicates, for example, that the junction is no longer operating as a *unilateral* circuit element—that is, a circuit element that conducts electrons in only one direction. Thus, an important test for the condition of the transistor is the reverse current through the collector-base junction. This test is illustrated in Fig. 10-1.

The reverse current, called I_{CBO}, is simply obtained by measuring the amount of current (in microamperes) between the collector and base when the junction is reverse biased. The parameter I_{CBO} is also referred to simply as I_{CO}.

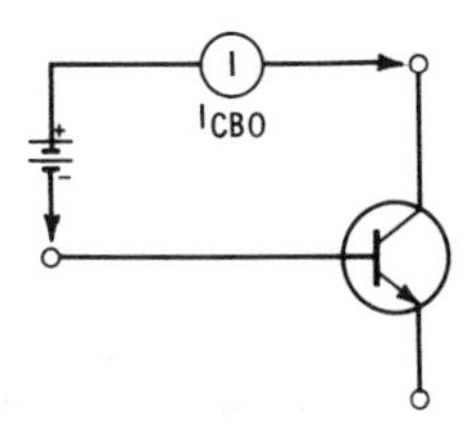

Fig. 10-1. Testing the collector-base junction for leakage.

The emitter-base junction of a transistor is also like a semiconductor diode. The difference between this junction and the collector-base junction being that this junction is normally forward biased. However, if it is operating correctly, there should be very little current when the junction is reverse biased. Again, the amount of current through it is an indication of the condition of the junction. Fig. 10-2 shows a test for measuring leakage in the emitter-base junction. The reverse-bias current is referred to as I_{ECS}, and it is measured with the collector and base leads shorted together. If the collector and base are not shorted together, and the same test is performd, the reverse-bias current is referred to as I_{EBO} or simply as I_{EO}.

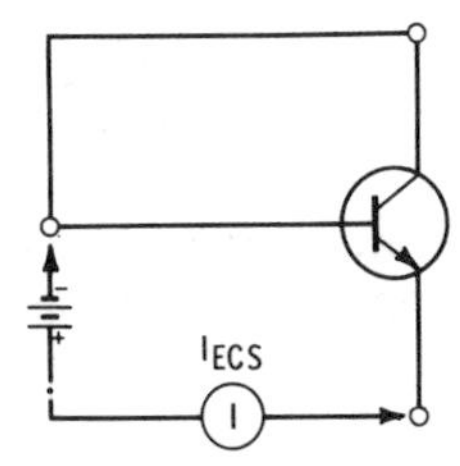

Fig. 10-2. Testing the emitter-base junction for leakage.

If the transistor is operating properly, there should be very little current between the emitter and collector when the base is open circuited. According to the theory of transistor operation, a forward bias on the emitter-base junction is necessary in order to obtain an appreciable collector current. An important test for the condition of the transistor is to measure the amount of emitter-collector current flow with the base open. This test is illustrated in Fig. 10-3. The current under this condition is called I_{CEO}.

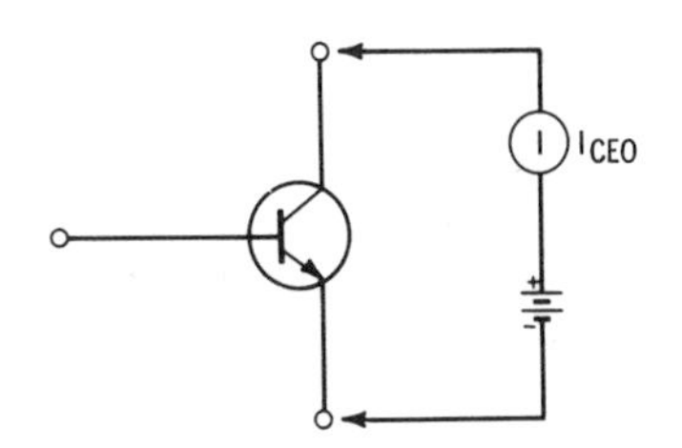

Fig. 10-3. Emitter-to-collector leakage with base open.

Commercially available transistor testers usually measure one or more of the parameters described for Figs. 10-1 through 10-3. An ohmmeter may be used for a quick test of the condition of the transistor. The ohmmeter test is illustrated in Fig. 10-4. To distinguish the ohmmeter leads in this illustration, positive and negative symbols are used. In practice, one of these leads will be positive and the other will be negative, depending upon the polarity of the battery within the ohmeter. It is important for a technician to know the polarity of his ohmmeter leads when troubleshooting in transistor circuits.

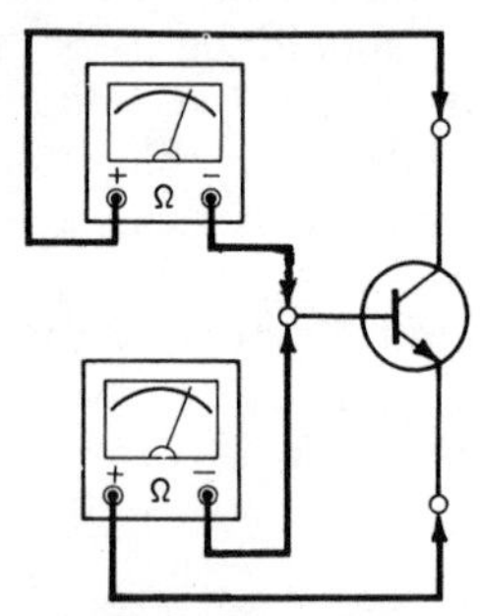

(A) *Reverse bias, high resistance.*

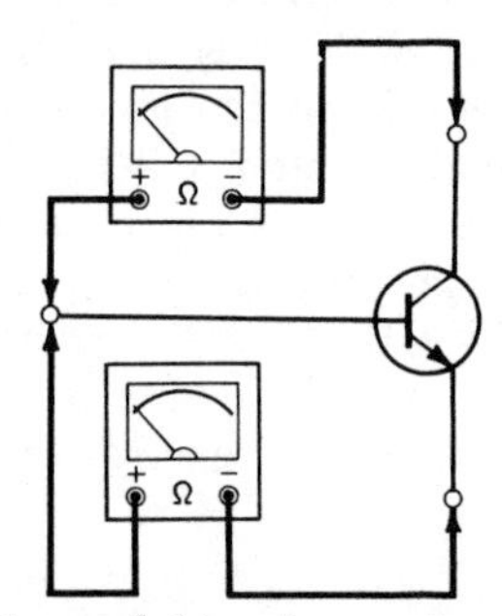

(B) *Forward bias, low resistance.*

Fig. 10-4. A simple ohmmeter test for a transistor.

The ohmmeter is connected as in Fig. 10-4A with the negative terminal applied to the base of the npn transistor, and the positive terminal of the ohmmeter applied first to the collector and then to the emitter terminals. A high resistance reading should be obtained for both measurements. When the ohmmeter leads are reversed as shown in Fig. 10-4B, both of the junctions in the transistor are forward biased and the resistance reading should be low for both measurements.

If the polarity of the ohmmeter is not known, or if it is not known whether the transistor is an npn or pnp type, then the same test is applied. You simply test for a high resistance reading when one ohmmeter lead is applied to the base, and a low resistance reading when the other ohmmeter lead is applied to the base.

It is best to make the test of Fig. 10-4 with the ohmmeter in the ×100 position, since a high reverse voltage from the ohmmeter circuit could damage a junction within the transistor.

Fig. 10-5 shows a test that is sometimes applied to a transistor. This particular test is important because an ohmmeter can be used for applying voltage between the collector and the emitter, provided that the ohmmeter leads are connected so that they forward bias the transistor for conduction. (For example, with an npn transistor the positive lead is applied to the collector and the negative lead is applied to the emitter.) In a standard test for obtaining the value of I_{CER}, a 10K or similar value of resistance is connected between the emitter and the base.

If you are using an ohmmeter across the collector-emitter junction, you can momentarily short circuit the emitter lead to the base lead. If the transistor is operating properly, this should indicate a decrease in collector current as indicated by an increase in the resistance measurement. Some technicians prefer to use a resistive probe rather than short circuiting the emitter-base junction to make this test. It is a quick test to see if a transistor will amplify in a circuit.

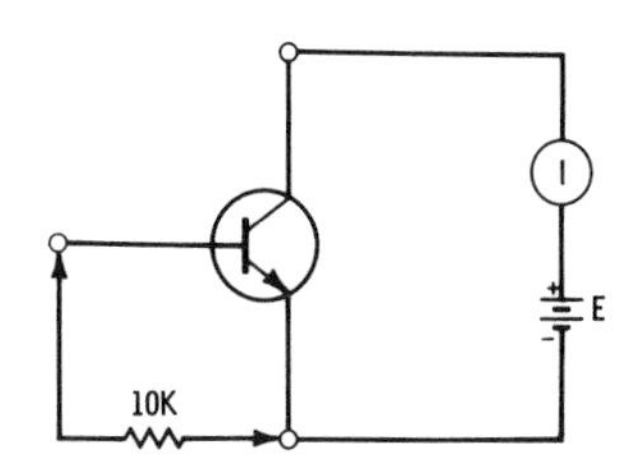

Fig. 10-5. Testing a transistor.

In an earlier chapter, the h_{FB} (dc alpha) was described as the dc collector current divided by the dc emitter current when the transistor is connected in a grounded-base configuration. The h_{FE} (dc beta) is the dc collector current divided by the dc base current when the transistor is connected in a grounded-emitter configuration. These dc parameters are not as significant as the h_{fb} (ac alpha) and the h_{fe} (ac beta). The value of h_{fb} is obtained when a small change in collector current is divided by the small change in emitter current that produced the collector current change (for the grounded-base configuration). The value of h_{fe} is obtained when a small change in collector current is divided by the small change in base current that produced the collector current change (for the grounded-emitter configuration). Fig. 10-6 shows how the voltage variations are obtained for determining the values of h_{fb} and h_{fe}. The delta symbol ($\triangle$) is used for indicating *small change.*

In-Circuit DC Voltage Checks

When it is not convenient to remove the transistor for testing, there are some dc voltage measurements that can be made in the circuit to determine if it is operating correctly.

Fig. 10-7 shows a typical amplifier circuit. The collector circuit is grounded through R3, and the positive voltage is applied to the emitter through R4. In another verision of this circuit, point "a" is grounded and a negative voltage is applied to the collector. In either case, since this is a pnp transistor, the voltage at the base

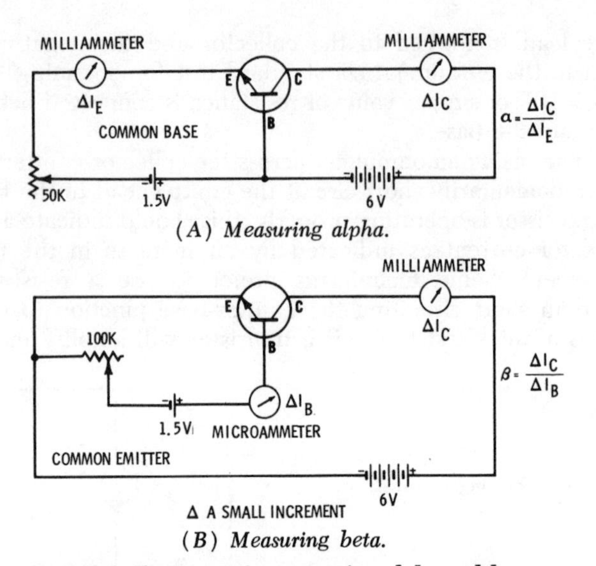

(A) *Measuring alpha.*

(B) *Measuring beta.*

Fig. 10-6. Test setup for measuring alpha and beta.

must be negative with respect to the emitter to forward bias that junction. Normally, we can expect to measure a base-emitter voltage of 0.15 to 0.7 volts, the higher value being for silicon types of transistors and the lower value for germanium types. Since the dc bias voltage is so small, it will usually be easier to measure directly between points "c" and "d" rather than measuring from ground to point "c" and from ground to point "d" and determining the difference.

If you have determined that the base bias is correct, the next step is to determine if there is collector current. This can be determined by a simple voltage measurement from point "b" to ground. If there

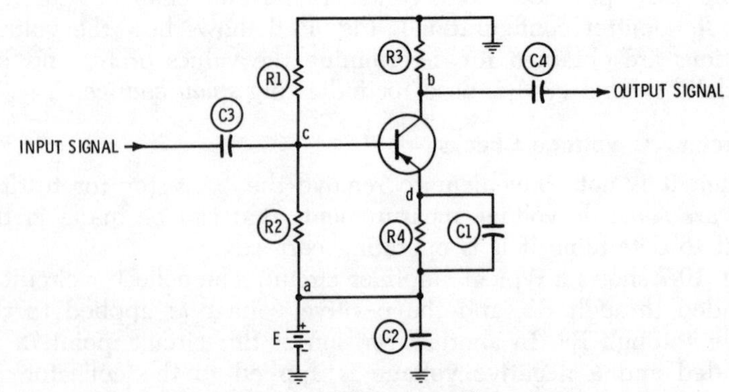

Fig. 10-7. Typical pnp amplifier circuit.

is a voltage drop across resistor R3, there is collector current through R3. If there is no voltage drop, then the transistor is probably open. Note that you cannot determine for sure if there is collector current by measuring a voltage across the emitter resistor. The base-to-emitter current is through this resistor, and this will produce a small voltage drop from point "a" to point "d." Of course, if you know the value of voltage that the emitter *should* have, and the emitter voltage is found to be incorrect, you should have a good idea whether or not there is collector current.

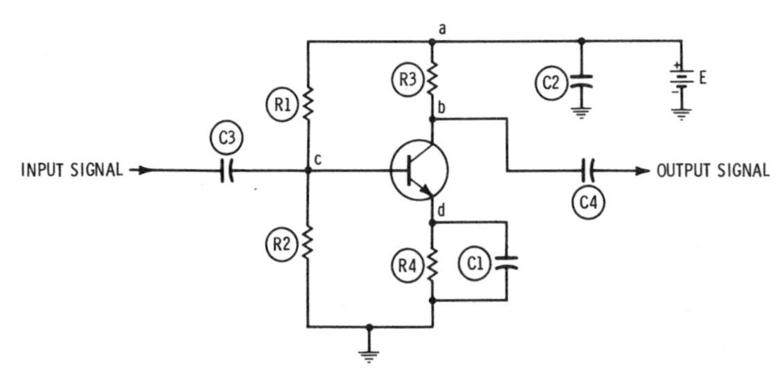

Fig. 10-8. An npn transistor amplifier circuit.

The circuit of Fig. 10-8 shows an npn transistor. The emitter current is returned to ground through R4, and the collector goes to the positive voltage source through R3. As before, a good starting point is to measure the dc bias voltage between points "c" and "d" and determine if the transistor is properly forward biased. As with the pnp transistor, the bias voltage should range from 0.15 to 0.7 volts.

If the base is properly biased and the transistor is properly conducting, there should be a voltage drop across resistor R3. This can be determined by measuring the voltage at point "a" and the voltage at point "b" with respect to ground, taking the difference. It can also be determined by measuring directly across the resistor if it is accessible.

Some technicians prefer to use a short between the emitter and base of the transistor to determine if the transistor is able to amplify. To do this, they momentarily short between points "c" and "d" while measuring the voltage at point "b" of the circuit in Fig. 10-7 or Fig. 10-8. When the emitter is shorted to base, the forward bias is removed and the collector current should decrease. For the circuit in Fig. 10-7, this will cause the voltage at point "b" to approach zero. For the circuit of Fig. 10-8, removing the bias will remove the collector current and cause the voltage at point "b" to approach the value of the supply voltage E.

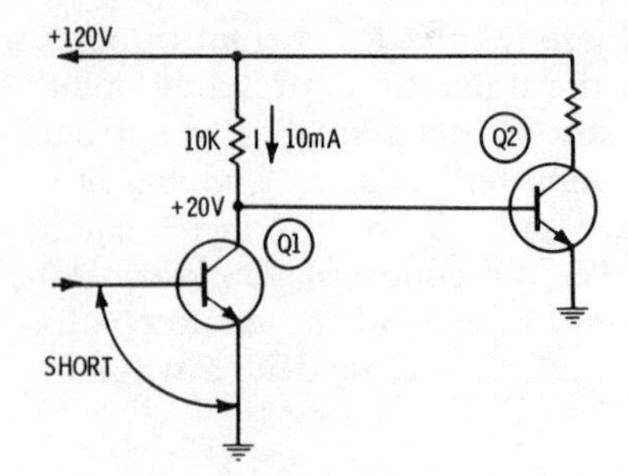

Fig. 10-9. An example of a circuit in which an emitter-to-base short should should not be used as a test.

Actually, any change in collector voltage produced by a change in bias voltage will be an indication that the transistor is capable of conducting and amplifying. Sometimes, the base may be grounded to produce the required change. However, it is not always a good idea to ground the base as an alternative to shorting it to the emitter. For example, in the circuit of Fig. 10-7 grounding point "c" will place the power-supply voltage directly across resistor R2. If this is a relatively small resistor, the power supply will be unnecessarily loaded. Furthermore, if capacitor C1 happens to be shorted, or if there is no emitter resistor for stabilization, then grounding point ""c" will place all of the supply voltage E across the emitter-base junction of the transistor and the base-to-emitter current will be excessive.

The technique of shorting the emitter to the base to determine if the transistor is operating is not always a safe one. Specifically note that in the circuit of Fig. 10-9, if the emitter-base leads are shorted on transistor Q1, its collector current will stop. This will cause the base voltage of Q2 to rise to an excessively high value and likely destroy the transistor. Furthermore, the rise in voltage at the collector of Q1 may be so great that the collector-to-base voltage will exceed the peak inverse voltage rating of that junction and transistor Q1 will be destroyed.

Fig. 10-10 shows a circuit in which shorting the emitter lead to the base lead will not produce noticeable results. This transistor is oper-

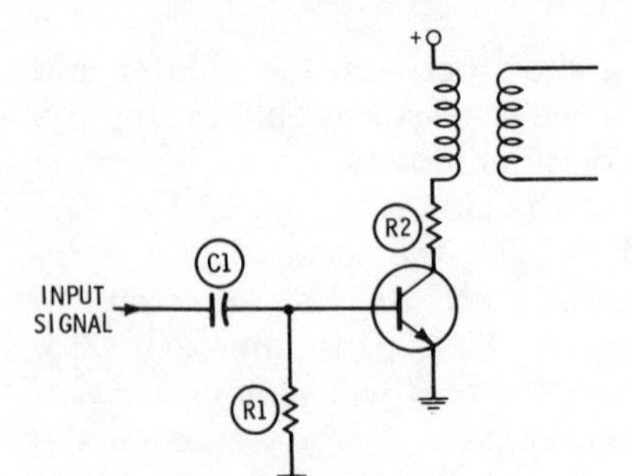

Fig. 10-10. An emitter-to-base short will not help to evaluate the transistor in this circuit.

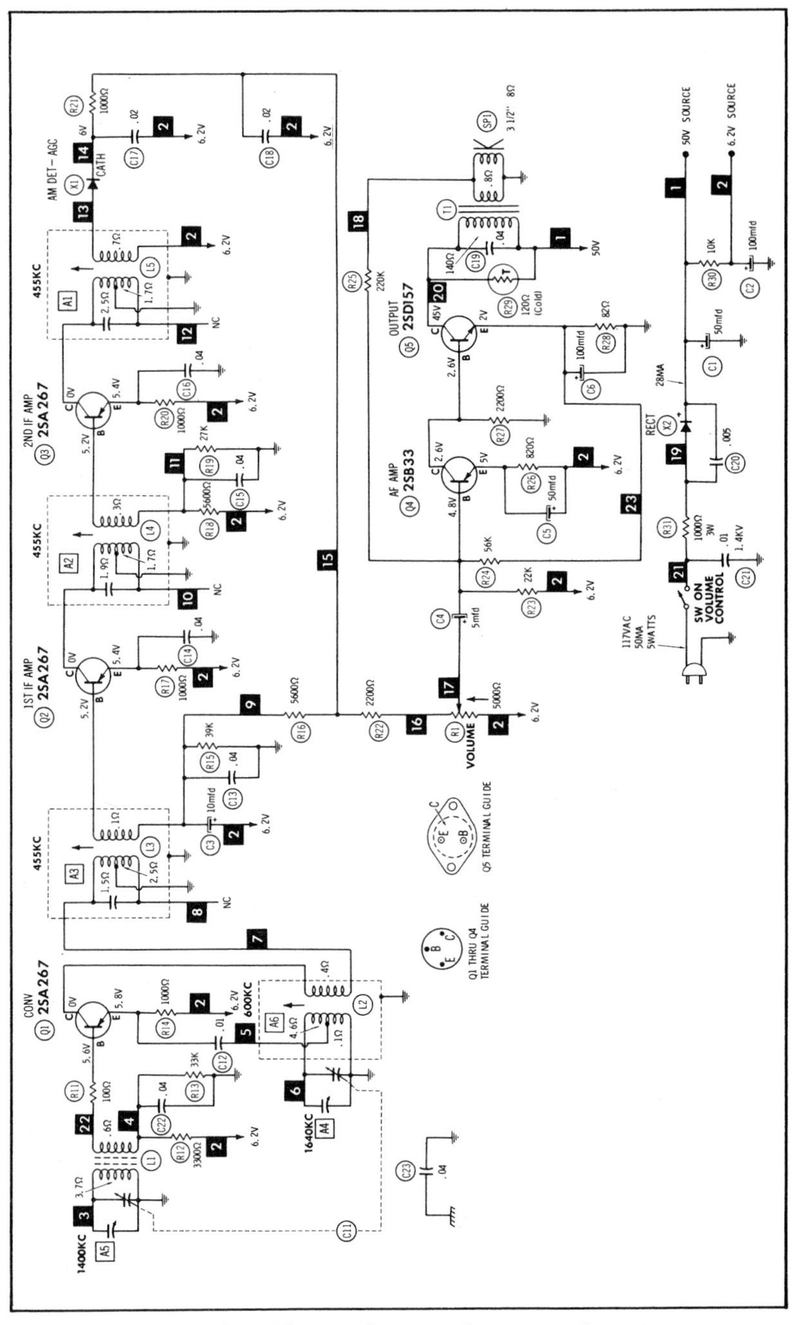

Fig. 10-11. Voltage values are shown on schematic.

ated with zero base bias voltage. In other words, the transistor is being operated Class B. It conducts only when there is an input signal voltage present. Therefore, grounding the emitter to the base will produce no change in collector voltage in this case.

A schematic diagram of the circuit being tested is a valuable aid to troubleshooting by dc voltage measurement. Note that in Fig. 10-11, all of the dc voltages for the transistors are shown. By checking the schematic diagram, you should be able to remove any doubt as to whether or not a transistor will be destroyed by shorting the emitter to the base for testing purposes.

If dc voltage measurements in a transistor circuit indicate that it is not conducting, and the bias voltage is normal, the transistor probably is open. If the transistor is not conducting, and if there is no base bias, it is likely that there is a circuit defect rather than a defective transistor. If the conduction of the transistor is too high and the bias is low or of the wrong polarity, a leaky transistor is a good possibility.

Precautions When Troubleshooting Transistorized Equipment

When troubleshooting in any kind of equipment, whether it is tube type or transistor type, it is important to use the proper probe. Chart 1 shows some of the more important probes for use in troubleshooting. The low-capacitance probe is usually preferred when troubleshooting with an oscilloscope in most types of equipment utilizing transistor circuits.

It is very important when connecting and disconnecting leads for making measurements, that you avoid short circuiting points in transistor circuits since this could be destructive to a transistor. Even though the circuit that you are measuring in has only low voltages, you should always de-energize the receiver when connecting and disconnecting test leads.

You should also have a good knowledge of the equipment that you are working with, including your test equipment, soldering iron, etc. Remember that ac leakage can introduce voltages in transistors which can exceed their junction breakdown rating. You should know your ohmmeter lead polarities, and you should also have knowledge of the amount of voltage that your ohmmeter places across a circuit during tests. Remember that an ohmmeter can damage a transistor with a voltage that exceeds the peak-inverse voltage rating or the forward-current limit of the transistor junctions.

Transient voltages can ruin transistors so you should avoid using such improper techniques as momentarily arcing the CRT anode lead to the chassis to determine if high voltage is present. This can easily ruin the transistors in the circuit, and also it can damage the flyback transformer.

When replacing transistors, diodes, or very small component parts, always replace the heat sink if one is used. Also, be careful not to overheat components or the printed circuit board by holding the iron on the connection too long.

Chart 10-1. VTVM and Oscilloscope Probes.

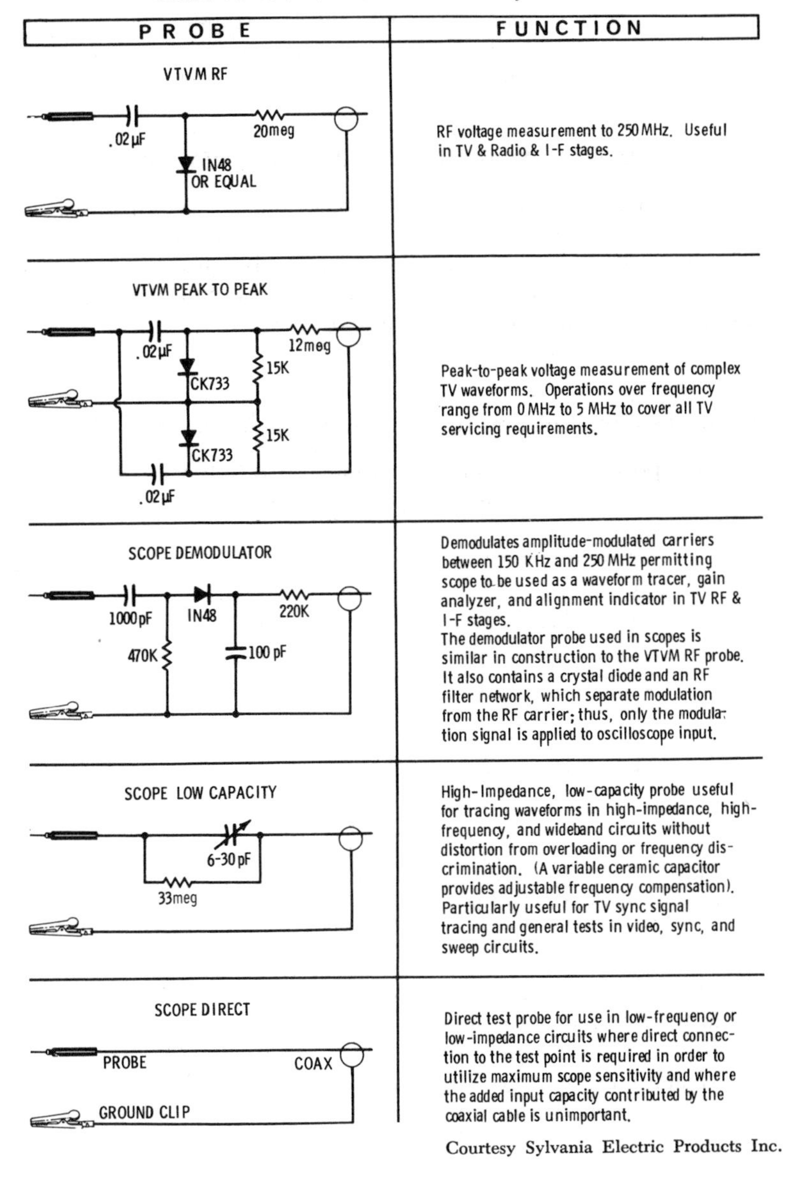

PROBE	FUNCTION
VTVM RF	RF voltage measurement to 250 MHz. Useful in TV & Radio & I-F stages.
VTVM PEAK TO PEAK	Peak-to-peak voltage measurement of complex TV waveforms. Operations over frequency range from 0 MHz to 5 MHz to cover all TV servicing requirements.
SCOPE DEMODULATOR	Demodulates amplitude-modulated carriers between 150 KHz and 250 MHz permitting scope to be used as a waveform tracer, gain analyzer, and alignment indicator in TV RF & I-F stages. The demodulator probe used in scopes is similar in construction to the VTVM RF probe. It also contains a crystal diode and an RF filter network, which separate modulation from the RF carrier; thus, only the modulation signal is applied to oscilloscope input.
SCOPE LOW CAPACITY	High-Impedance, low-capacity probe useful for tracing waveforms in high-impedance, high-frequency, and wideband circuits without distortion from overloading or frequency discrimination. (A variable ceramic capacitor provides adjustable frequency compensation). Particularly useful for TV sync signal tracing and general tests in video, sync, and sweep circuits.
SCOPE DIRECT	Direct test probe for use in low-frequency or low-impedance circuits where direct connection to the test point is required in order to utilize maximum scope sensitivity and where the added input capacity contributed by the coaxial cable is unimportant.

Courtesy Sylvania Electric Products Inc.

When replacing a power transistor, be sure to use the silicon grease on both sides of the mica insulator, if one is used, so that you have a good thermal connection.

You should never operate a transistor receiver with the speaker disconnected because of the possibility of destroying the power output transistors.

Last, but not least, remember that transistor stages are often direct coupled, and an incorrect dc voltage in one circuit can cause a malfunction in another circuit.

PROGRAMMED QUESTIONS AND ANSWERS

Starting with question number 1, select the answer that you feel is correct. If you feel that (A) is correct, proceed to block number 17 as directed. If you feel that (B) is correct, proceed to block number 9 as directed. If you feel that more than one answer is correct, choose the one that you think is the *most* correct.

1 In the circuit of Fig. 10-12, point X should be:

(A) positive with respect to ground. (Go to block number 17.)
(B) negative with respect to ground. (Go to block number 9.)

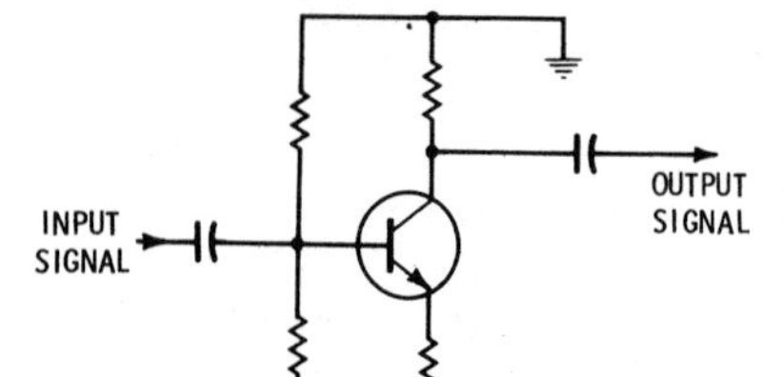

Fig. 10-12. A Class-A amplifier circuit.

2 Your answer is wrong. Study Chart 1 regarding the use of the direct probe. Go to block number 20.

3 The collector current will decrease as indicated by an increase in the ohmmeter reading.

Here is the next question . . .

Which will be larger—the forward bias on a Class-B transistor amplifier; or, the forward bias on a Class-A transistor amplifier? (Go to block number 7.)

. . . .

4 Your answer is wrong. The voltage across the primary of the transformer is usually negligible and it is doubtful if you would measure a change. In any case, cutting the transistor off would remove all of the drop across the load and the collector would become more negative, not positive. Go to block number 11.

. . . .

5 Your answer is correct. Note the typical resistance values listed for a low forward resistance and a high reverse resistance. These values are correct.

Here is the next question . . .

When testing a transistor with an ohmmeter, set the ohmmeter multiplier switch on:

(A) R × 1 meg. (Go to block number 23.)

(B) R × 100. (Go to block number 14.)

. . . .

6 Your answer is wrong. If Q5 is open at the base, transistor Q4 will still conduct through R27. Go to block number 15.

. . . .

7 The forward bias on a Class-A transistor amplifier will be larger.

Here is the next question . . .

Which type of probe will you use for looking at the output signal of a transistor video amplifier? (Go to block number 26.)

. . . .

8 Your answer is wrong. If R27 is open, the collector will be positive with respect to ground, but the collector voltage will be high. Go to block number 15.

. . . .

9 If point X is negative with respect to ground, and the collector is grounded through the collector resistor, then the collector is positive with respect to the emitter.

Here is the next question . . .

The forward resistance on the emitter-base junction of a transistor is about 220 ohms, and the reverse resistance is about 20K ohms. Which of the following statements is correct?

(A) The transistor is defective. (Go to block number 24.)

(B) As far as this junction is concerned, the transistor is all right. (Go to block number 13.)

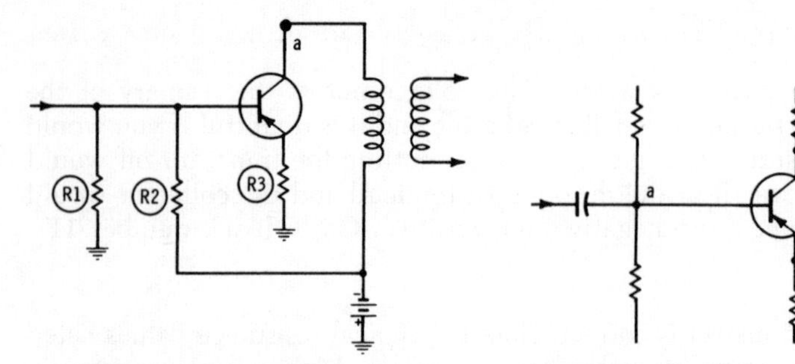

Fig. 10-13. A pnp amplifier circuit. Measure voltage at point "a."

Fig. 10-14. The bias is measured between points "a" and "b."

. . . .

10 You do not have the complete circuit, but the bias is typical for a pnp transistor.

Here is the next question . . .

You are using an ohmmeter to test a pnp transistor. You have the ohmmeter positive lead on the collector, and the ohmmeter negative lead on the base. Assuming that the transistor is good, the ohmmeter should show:

(A) a low resistance (about 220 ohms). (Go to block number 5.)

(B) a high resistance (about 150K ohms). (Go to block number 25.)

. . . .

11 There will be no appreciable change in the voltage at point "a" in this case. This is an example of a circuit in which shorting across the emitter-base junction does not show a change in collector voltage. You could, however, measure the emitter voltage to see if a change occurs.

Here is the next question . . .

In the amplifier circuit of Fig. 10-14, the voltage at point "a" is +9.4 volts with respect to ground, and the voltage at point "b" is +9.6 volts with respect to ground. Which of the following is true?

(A) The polarity of the bias voltage is wrong. (Go to block number 21.)

(B) The amount of bias voltage is wrong. (Go to block number 16.)

(C) The bias is probably correct for this amplifier. (Go to block number 10.)

. . . .

12 Your answer is wrong. The collector current will decrease, but this will be indicated by an *increase* in the resistance reading on the ohmmeter. Go to block number 3.

. . . .

13 These are typical values for forward and reverse resistance of the emitter-base junction.

Here is the next question . . .

For the circuit of Fig. 10-11, you wish to use an oscilloscope to observe the envelope of the waveform on the base of Q2. You should use a:

(A) direct probe. (Go to block number 2.)
(B) demodulator probe. (Go to block number 20.)
(C) low-capacity probe. (Go to block number 18.)

. . . .

14 The R × 100 scale is favored for measuring across transistor junctions.

Here is the next question . . .

An ohmmeter is connected across the emitter and collector leads of a good transistor. The polarity of the ohmmeter is such that collector current will flow. When the emitter and base leads are connected, the resistance reading on the ohmmeter will:

(A) increase. (Go to block number 3.)
(B) decrease. (Go to block number 12.)
(C) not change. (Go to block number 19.)

. . . .

15 Your answer is right. If the collector of Q4 is open, there is no conduction path through R27.

Here is the next question . . .

In the circuit of Fig. 10-13 you momentarily short across R1 while measuring the voltage at point "a." Which of the following is correct?

(A) You will destroy the transistor. (Go to block number 22.)
(B) There will be no appreciable change in the voltage at point "a" (Go to block number 11.)
(C) The voltage at point "a" should become slightly more positive. (Go to block number 4.)

. . . .

16 Your answer is wrong. The forward bias is 0.2 volts, which is within the typical range of values. Go to block number 10.

. . . .

17 Your answer is wrong. The emitter should be negative with respect to the base and collector in an npn transistor. Go to block number 9.

. . . .

18 Your answer is wrong. Study Chart 1, regarding the use of probes. Go to block number 20.

. . . .

19 Your answer is wrong. Shorting the emitter to the base puts a reverse bias on the emitter-base junction and increases the resistance to current flow. Go to block number 3.

. . . .

20 A demodulator probe will permit you to observe the envelope of the waveform of the i-f signal. Of course, the signal will be rectified, so you will only see the positive or negative half of the envelope.

Here is the next question . . .

In the circuit of Fig. 10-11, the voltage at the collector of Q4 measures zero volts, and the base voltage measures 4.8 volts. Which of the following is the most likely cause?

(A) R27 is open. (Go to block number 8.)
(B) Q5 is open at the base. (Go to block number 6.)
(C) The collector lead of Q4 is open. (Go to block number 15.)

. . . .

21 Your answer is wrong. The voltages show that the base is negative with respect to the emitter as required for forward biasing the pnp transistor. Go to block number 10.

. . . .

22 Your answer is wrong. Shorting across R1 removes the negative (forward) bias on the transistor and cuts it off. A transistor is not destroyed by operating it at cutoff. Go to block number 11.

. . . .

23 Your answer is wrong. You should not measure across transistor junctions with the ohmmeter set on a high resistance scale. Go to block number 14.

. . . .

24 Your answer is wrong. Go to block number 13.

. . . .

25 Your answer is wrong. The ohmmeter is forward biasing the collector junction. This means that a low resistance reading should be obtained. Go to block number 5.

. . . .

26 You should use a low-capacity probe.
You have now completed the programmed questions and answers.

. . . .

PRACTICE TEST

1. When using a scope for looking at the video amplifier signal, you should use a:

(a) low-capacity probe.
(b) demodulator probe.
(c) direct probe.
(d) voltage-doubler probe.

2. The bias on a certain transistor measures too low, but its collector current is too high. Which of the following is the likely cause?

(a) An open voltage divider circuit for biasing the base.
(b) A leaky transistor.
(c) An open emitter resistor.
(d) An open bypass capacitor in the emitter stage.

3. In a certain transistor circuit the dc voltage measurement indicates that the transistor is not conducting. The base bias voltage is measured and found to be correct. Which of the following is more likely to be true?

(a) The power supply is not operating.
(b) The transistor is open.
(c) The transistor is shorted.
(d) The power supply is at fault.

4. Which of the following is correct?

(a) $h_{fe} = \frac{\Delta I_c}{\Delta I_b}$.
(b) $h_{FE} = \frac{\Delta I_c}{\Delta I_b}$.

5, The emitter-collector terminals of a transistor have a voltage applied for conduction. At the same time, a 10K resistor is connected between the base and emitter. The collector current measured in this test is:

(a) I_{CER}.
(b) I_{CO}.
(c) I_{CES}.
(d) I_{EBR}.

6. In the normal operation of a transistor, which junction is reverse biased?

(a) The emitter-base junction.
(b) The collector-base junction.

7. The reverse current flow in the collector-base junction, with the emitter open, is called:

(a) I_{CBO}.
(b) I_{ECS}.
(c) I_{CEO}.
(d) I_{CER}.

8. Which of the following statements is true?

(a) The forward bias on a transistor that is operating Class A will be greater than the forward bias on a transistor that is operating Class B.
(b) The forward bias on a transistor that is operating Class B will be greater than the forward bias on a transistor that is operating Class A.

9. Which of the following is true?

(a) For an effective in-circuit test of a transistor, short the collector to the base.
(b) Shorting the base to the collector is almost sure to destroy the transistor.
(c) Shorting the base to the emitter is almost sure to destroy the transistor if it is in a Class-A amplifier circuit.
(d) Doubling the collector voltage for two or three minutes will rejuvenate a transistor with low conduction.

10. The base of a pnp transistor in a Class-A amplifier circuit has a voltage of +6.8. Which of the following voltages would you expect to measure on the emitter?

(a) +6.4 V.
(b) +9 V.
(c) +3 V.
(d) +7.2 V.

11. An ohmmeter is placed across the emitter-collector terminals of a transistor. When the collector and base leads are shorted together, the ohmmeter shows a lower resistance. Which of the following is true?

(a) The transistor is shorted.
(b) The emitter-base junction of the transistor is open.
(c) This is normal for a good transistor.
(d) This can never happen.

12. The collector circuit of a certain npn transistor is connected to ground through a transformer winding. You would expect the emitter of this transistor to be:

(a) positive with respect to ground.
(b) negative with respect to ground.

13. You are measuring the current drawn by a transistor radio by connecting a milliammeter across the ON-OFF switch. Which of the following is true?

(a) The switch must be in the ON position.
(b) If the radio has a Class-B audio amplifier, the current drawn will increase when the radio is tuned to a station.
(c) You cannot measure the current this way.
(d) If the radio has a Class-B audio amplifier, the current drain will decrease when the radio is tuned to a station.

14. In which of the following classes of amplifiers is the forward bias on the base of the transistor highest?

(a) Class A.
(b) Class B.
(c) Class C.

15. The forward emitter-base resistance on a certain transistor is 220 ohms, and the reverse resistance is 500 ohms. Which of the following is correct?

(a) The ratio is too low; the transistor is not good.
(b) The ratio is too high; the transistor is not good.
(c) As far as the transistor emitter-base junction is concerned, these are normal values of resistance.

16. To determine the reverse resistance of a pnp collector-base junction,

(a) place the positive lead of the ohmmeter on the collector terminal and the negative lead on the base terminal.
(b) place the negative lead of the ohmmeter on the collector terminal and the positive lead on the base terminal.

17. To check a transistor with an ohmmeter, set the ohmmeter on the:

(a) R × 100 position.
(b) R × 1 meg position.

18. For a three-gun color picture tube, the blue gun is mounted:

(a) either above or below the central axis of the tube.
(b) to the right of the central axis as viewed facing the screen.
(c) to the left of the central axis as viewed facing the screen.

19. Which of the following is not correct?

(a) The high-voltage section of a color television receiver must produce a constant-voltage output.
(b) During the setup procedure, the brightness control should be advanced beyond the normal range.
(c) If a conventional monochrome test pattern is not available, a cross-hatch generator can be used for checking linearity and height controls.

20. Which of these adjustments should be made first in the setup procedure?

(a) Purity.
(b) Picture size.
(c) Focus.
(d) High voltage.

21. Which of the following adjustments should be performed first in the setup procedure?

(a) Static convergence.
(b) Color purity.

22. Which of the following is not used for static convergence?

(a) Beam-positioning magnets.
(b) Lateral-correction magnet.
(c) Purity magnets.

23. When static convergence adjustments are made, the dots in the center should be:

(a) red.
(b) green.
(c) blue.
(d) white.

24. Which of the following is not adjusted or positioned to obtain optimum color purity?

(a) The color purity coil or magnet.
(b) The deflection yoke.
(c) The field neutralizing coil or magnet.
(d) The lateral correction magnet.

25. Which of the following would be a logical first step in obtaining color purity?

(a) Set the field neutralizing magnets (if used) in the neutral position.
(b) With the blue and green guns off, adjust purity magnet for an area of pure red illumination at the center of the screen (yoke moved back).
(c) With the blue and green guns on, adjust purity magnet for an area of pure red illumination at the center of the screen (yoke moved back).

26. Which of the following is the last step in a color-TV setup procedure?

(a) High-voltage adjustment.
(b) Gray-scale adjustment.
(c) Purity adjustment.
(d) Static convergence.

27. Which of the following is not aligned in a color receiver?

(a) The rf section.
(b) The i-f section.
(c) The luminance section.
(d) The bandpass amplifier.

28. The frequency response of the bandpass amplifier must be flat for a range of:

(a) 0.5 MHz above the color subcarrier and 4.5 MHz below the color subcarrier.
(b) 0.5 MHz on both sides of the color subcarrier.
(c) 4.5 MHz on both sides of the color subcarrier.
(d) 3.58 MHz on both sides of the color subcarrier.

29. When the correct flesh tones are obtained, the hue control should be:

(a) at its maximum counterclockwise position.
(b) in its midrange position.
(c) in its maximum clockwise position.

30. For the purpose of aligning the color section, the color-bar generator output terminals are connected to the:

(a) antenna terminals.
(b) color picture tube.
(c) luminance amplifier after the delay line.
(d) frequency control for the color subcarrier generator.

31. On a normal pattern obtained with a keyed rainbow generator:

(a) blue is on the left of the screen and red is on the right.
(b) yellow-orange is on the left of the screen and green is on the right.
(c) green is between the red and blue display.
(d) the vertical dark stripes should show the complete range of gray scale.

32. As a first step in troubleshooting a color television receiver:

(a) determine if the brightness control will operate throughout its range.
(b) determine if the horizontal and vertical hold controls are working.
(c) analyze the picture to determine if the trouble is in the monochrome or color section.
(d) measure the high voltage.

33. In a certain color receiver there is no monochrome picture when a monochrome signal is being received. When a color bar generator is used, the colors appear on the screen, but they do not have the proper brightness. The trouble is:

(a) in the video detector stage.
(b) in the luminance section between the color takeoff point and the picture tube.
(c) in the color sync section.
(d) due to improper convergence.

34. Which of the following would not be a likely cause of a color receiver not being able to reproduce shades of gray?

(a) One of the guns of the color picture tube being defective.
(b) The circuits which control the voltages applied to the picture tube not being adjusted correctly.
(c) One of the transistors in the luminance channel being defective.
(d) The color sync section being improperly aligned.

35. Green and magenta bars appear on the screen. This could be due to:

(a) hum modulating the green gun.
(b) hum modulating the blue and red guns.
(c) an open video detector.
(d) an open green gun.

36. If the color receiver will not reproduce any color when receiving a color signal, a logical place to start is:

(a) at the output of the delay line.
(b) at the picture tube.
(c) in the color oscillator and demodulator stages.
(d) in the matrix.

37. When signals from the color demodulators are lacking in amplitude, the result is:

(a) improper hues.
(b) loss of gray scale.
(c) desaturated colors.
(d) loss of brightness.

38. During reception of a monochrome signal, the color killer cuts off the:

(a) color subcarrier oscillator.
(b) video amplifiers.
(c) bandpass amplifier.
(d) matrix section.

39. Wrong colors are not likely to be caused by:

(a) improper alignment of amplifiers in the color sync circuit.
(b) improper adjustment of the hue control.
(c) improper voltages on the reactance tube across the 3.58-MHz oscillator.
(d) an incorrectly adjusted contrast control.

40. Diagonal stripes of variegated colors are most likely indicated by:

(a) a loss of color synchronization.
(b) a defective video amplifier stage.
(c) incorrect receiver agc voltage.
(d) a defective rf amplifier.

41. If the converter stage in a transistor radio receiver is operating properly, the base voltage:

(a) will not change when the receiver is tuned to a strong station.
(b) will change when the receiver is tuned to a strong station.

42. The cathode resistor in a pentode amplifier stage is open, but the plate and screen are both positive. A VOM will show:

(a) a positive voltage on the cathode.
(b) no voltage on the cathode.
(c) a negative voltage on the cathode.

43. A transistor junction may be damaged by excessive current flow when an ohmmeter is used to measure the forward resistance, if the ohmmeter is in the:

(a) R × 1 position.
(b) R × 100 position.
(c) R × 10K position.

44. When soldering parts into radios, use:

(a) 60-40 solder.
(b) 40-60 solder.
(c) pure tin.
(d) pure lead.

45. An ohmmeter is placed across an npn transistor with its negative probe connected to the emitter and its positive probe connected to the collector. With the ohmmeter thus connected, a 500K-ohm resistor is connected from collector to base. This will:

(a) decrease the resistance reading.
(b) increase the resistance reading.

46. Connection to the collector of a power transistor is:

(a) through the case or through a threaded stud.
(b) through the smallest terminal lead.

47. Low battery voltage in a transistor radio receiver is not likely to cause:

(a) distorted sound.
(b) weak sound.
(c) motorboating.
(d) static.

48. A certain radio that operates from the power line has a loud noise. The volume of this noise can be controlled by the volume control. This means that the:

(a) noise is due to a bad filter in the power supply.
(b) source of the noise must be ahead of the detector stage—that is, between the antenna and audio detector.

49. When troubleshooting a transistor television receiver:

(a) always disconnect the yoke.
(b) never operate the receiver with the yoke disconnected.
(c) always disconnect the speaker.

50. The emitter-base bias:

(a) is usually somewhat higher for germanium transistors than it is for silicon types.
(b) is always identical for germanium and silicon transistor types.
(c) is usually somewhat lower for germanium transistors than it is for silicon types.
(d) should always be such that the base is negative with respect to the emitter.

11

AGC, Power Supplies, and Waveform Analysis

KEYED STUDY ASSIGNMENTS

Howard W. Sams Photofact Television Course
Chapter 4—Power Supplies
Chapter 16—Automatic Gain Control

IMPORTANT CONCEPTS

Response Curves

Sweep generators (in conjunction with oscilloscopes and marker generators) are used for obtaining response curves of the tuner, the video i-f section, and the color bandpass amplifiers.

Fig. 11-1 shows the test setup for obtaining a tuner response curve. A 1.5-volt battery is used in place of the tuner agc voltage so that the response will not vary with changes in the amplitude of the signal at the video detector. In this case, a negative voltage is used for the agc, but in transistor receivers it is possible that a positive voltage will be used.

The input signal from the sweep generator is fed to the antenna terminals, and the oscilloscope is connected to the *looker point*, which is a test point at the output of the mixer stage.

Fig. 11-2 shows a typical tuner response curve. The curve should be relatively flat over a 6-MHz range for each channel being tuned. Normally, the sweep width is set for something greater than 6 MHz

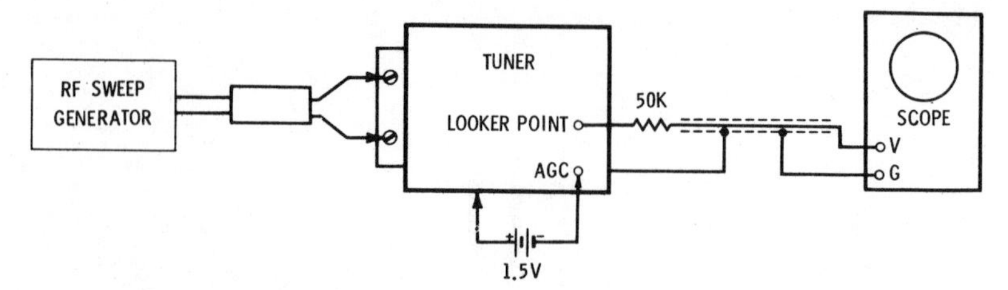

Fig. 11-1. Using a sweep generator to check tuner response.

(8 or 10 MHz) so that the curve goes well beyond the required limits of flat response. Markers, which will be discussed later, may be used to indicate specific frequencies on the response curve.

Fig. 11-3 shows how a response curve is obtained for the video i-f section of a television receiver. The sweep generator is fed into the input of the i-f stages, and the output signal for the scope vertical input is taken from across the video-detector load. The internal horizontal sweep of the oscilloscope is turned off, and the horizontal sweep is provided by the sweep generator.

A marker oscillator is used in the test setup of Fig. 11-3. Markers are used to identify a certain frequency on the response curve. A disadvantage of the test setup shown in Fig. 11-3 is that the signal from the marker oscillator must pass through the i-f stages along with the sweep signal. It is possible for the marker signal to distort the resulting curve and give a false indication of the i-f amplifier response.

A much preferred method of obtaining an i-f response curve is shown in Fig. 11-4. This is called the *post marking system.* The marker generator is fed to a *marker adder,* which also receives the response curve from the video detector of the TV receiver. Thus, the markers are added *after* the signal has passed through the receiver, and therefore, the marker signal need not pass through the tuned circuits in the i-f stages. The output of the marker adder then is delivered to the oscilloscope. It is interesting to note that, with the post marker setup, the markers are visible on the sweep even in the

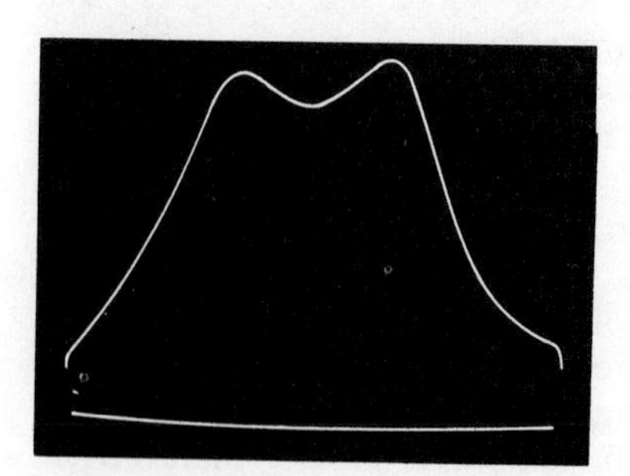

Fig. 11-2. A typical tuner response curve.

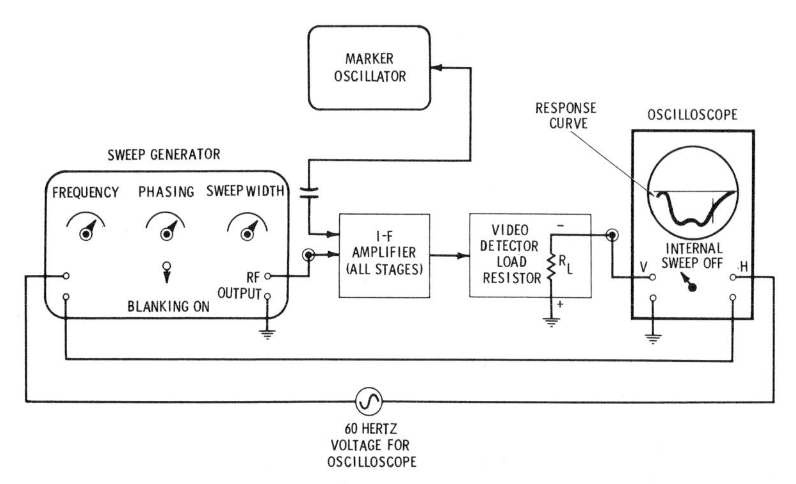

Fig. 11-3. Test setup used for obtaining the i-f response curve.

event that the output from the detector is lost. This is not true of the test setup shown in Fig. 11-3.

The disadvantage of the marker adder system is that an additional piece of test equipment is needed for obtaining the response curve. If a marker adder is not used, it is very important to keep the amplitude of the marker signals as low as possible to avoid unnecessary distortion of the response curve.

Fig. 11-5 shows an ideal i-f response curve. You will note that the video i-f carrier is above the sound i-f carrier, although the video signal is transmitted below the sound carrier. This is because the signal is reversed when it goes through the mixer stage.

The sloping response of the video carrier is needed in order to compensate for the fact that the video is sent as a vestigial sideband

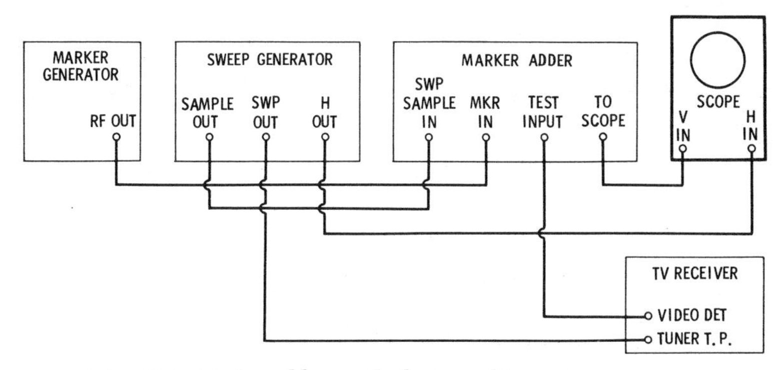

Fig. 11-4. Marker-adder method of marking response curve.

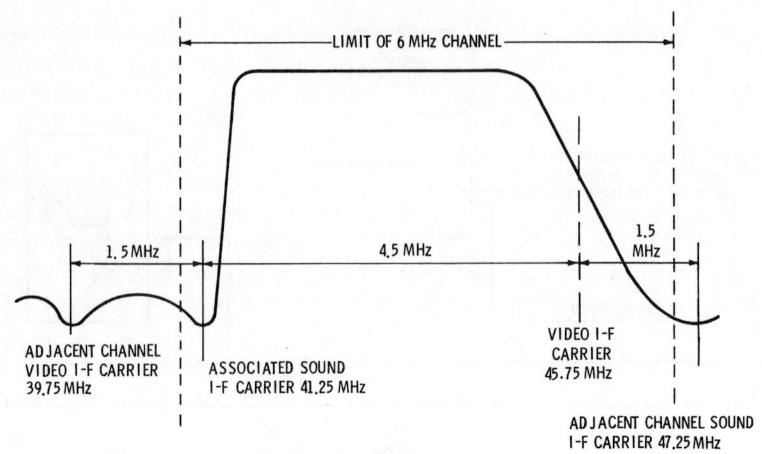

Fig. 11-5. Ideal i-f response curve.

signal. The receiver response is reduced at the video carrier because both sidebands are transmitted in that region of the signal, and the slope prevents overemphasis of the lower video frequencies.

Fig. 11-6 shows some typical i-f response curves with markers. In Fig. 11-6A the markers are vertical, and in Fig. 11-6B the markers are horizontal. The advantage of having the markers horizontal is that they cross the response curve at a right angle and give a clearer indication of the exact marker point. Some types of test equipment provide a choice of vertical or horizontal markers.

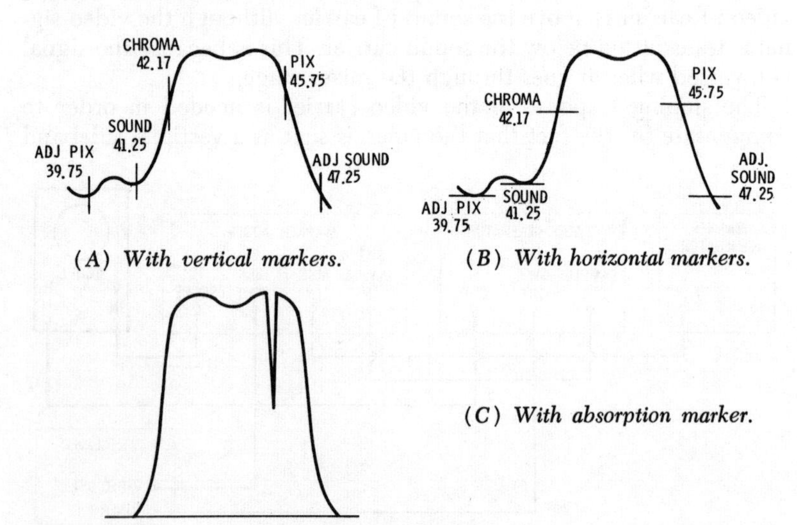

(*A*) *With vertical markers.*

(*B*) *With horizontal markers.*

(*C*) *With absorption marker.*

Fig. 11-6. Typical i-f response curves.

The absorption marker shown in Fig. 11-6C is obtained by using a tuned circuit to produce a notch in the response curve. The tuned circuit is arranged so that the point being marked will not pass through to the oscilloscope. This represents a loss of response at that point and produces the notch.

The color bandpass amplifier of the receiver should be able to pass about 0.5 MHz of signal on either side of the 3.58-MHz subcarrier. Fig. 11-7 shows two methods of obtaining bandpass response curves. Fig. 11-7A shows how the response curve for the bandpass amplifier can be obtained with a video sweep generator and marker. As

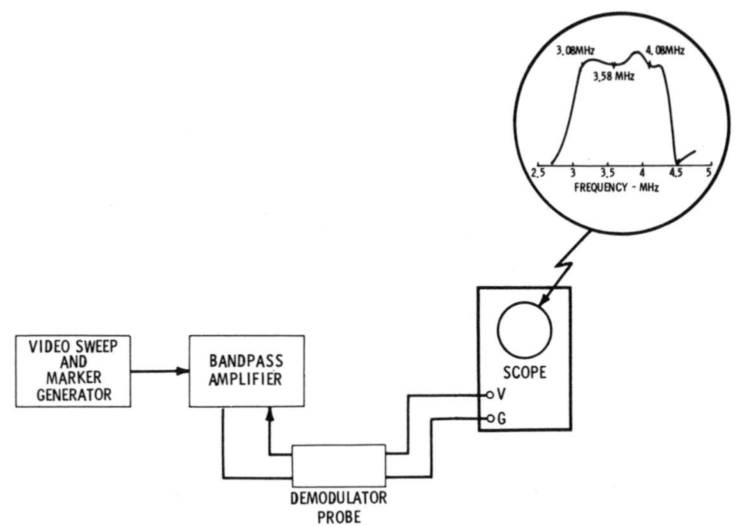

(A) *Standard sweep and marker generator setup.*

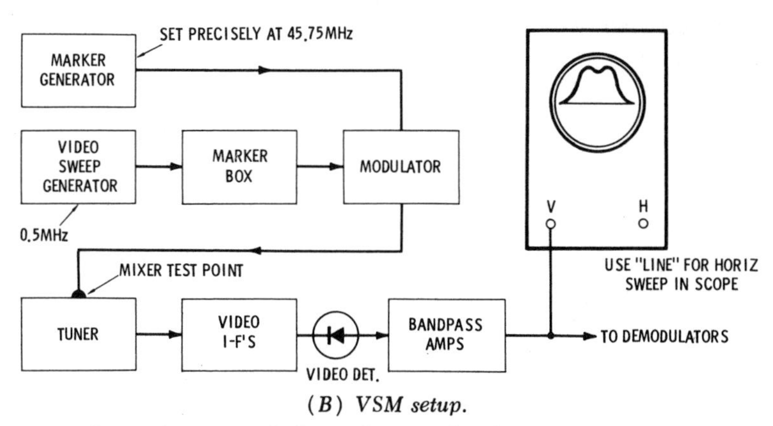

(B) *VSM setup.*

Fig. 11-7. Two methods of obtaining bandpass response curve.

mentioned before, a marker adder system may be used so that the marker signals do not have to pass through the bandpass amplifier.

It is sometimes difficult to obtain the response curve with the test setup shown in Fig. 11-7A because some sweep generators cannot be adjusted to sweep a 2-MHz range around the 3.58-MHz point. If the bandpass-amplifier response curve cannot be obtained in this manner, the test setup shown in Fig. 11-7B can be used. This is called the *VSM* method. (The initials VSM stand for Video Sweep Modulator.) In this setup the signal from the video sweep generator sweeps from 0 to 5 MHz. The sweep signal is marked at 3.08, 3.58, 4.08, and 4.5 MHz. The sweep signal is fed to a modulator where it is mixed with a signal from the marker generator set at precisely 45.75 MHz. The two signals are combined in the modulator and fed through the video i-f stages to the bandpass amplifier.

By using an absorption marker in the VSM test setup, signals to be marked are actually removed from the composite signal. Therefore, it is not necessary to feed additional marking signals through the video i-f stage. It is important to remember that with the VSM method, the sweep signal must pass through the video i-f stages as well as the bandpass amplifier. If the video i-f stage is out of alignment, it will not be possible to obtain the bandpass amplifier response curve. It is a good idea, therefore, to check the video i-f response in a test setup.

Using a Vectorscope

Before discussing the vectorscope setup, we will first review the method of generating Lissajous patterns on the oscilloscope.

Lissajous patterns are used for comparing the frequency or the phase of two signals. Fig. 11-8 shows what happens when two sine waves of the same frequency, but 90° out of phase, are applied to the horizontal and vertical inputs of the oscilloscope. The resulting Lissajous pattern is a circle generated by the two waveforms. The corresponding points on waveforms "A" and "B" are marked on the Lissajous pattern.

As shown in Fig. 11-9, when the two sine waveforms are in phase, a straight line pattern is obtained. This pattern is still referred to as being a Lissajous pattern.

If the input sine waves are 180° out of phase, a straight line will be produced again, but it will slant downward rather than upward as shown in Fig. 11-9.

Lissajous patterns can be used for troubleshooting. For example, in testing an amplifier for phase distortion, a sine-wave signal is applied to the input of the amplifier, and the same signal is applied to the horizontal input of the oscilloscope. The output waveform from the amplifier is applied to the vertical input of the oscilloscope.

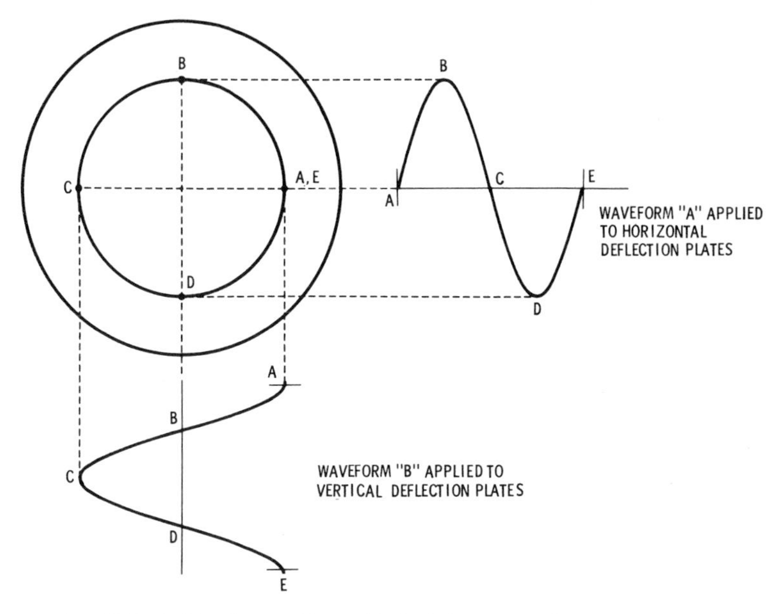

Fig. 11-8. A Lissajous pattern obtained by comparing waveforms on an oscilloscope.

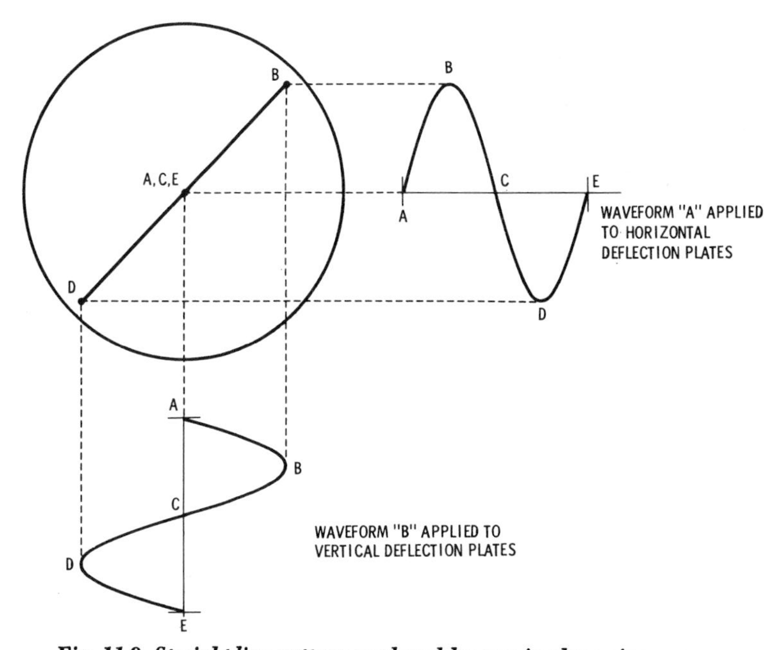

Fig. 11-9. Straight-line pattern produced by two in-phase sine waves.

The oscilloscope is adjusted so that each signal produces a two inch deflection. If a straight line is produced, there is no phase shift distortion in the amplifier. However, if the pattern is an ellipse, further investigation in the amplifier is indicated. The greater the phase shift distortion, the greater the opening of the ellipse.

The vectorscope produces a Lissajous pattern by comparing the R – Y and B – Y signals on the CRT. The R – Y signal is applied to the vertical deflection system and the B – Y signal is applied to the horizontal deflection system. These signals are obtained when a color-bar generator is connected to a receiver in a normal fashion.

Fig. 11-10 shows vectorscope patterns. Fig. 11-10A shows a theoretical daisy pattern obtained on the vectorscope by comparing the R – Y and B – Y signals. Note that there are ten petals on the daisy corresponding to the ten different signals produced by the color-bar generator. Fig. 11-10B shows an actual daisy pattern for comparison with the theoretical one of Fig. 11-10A.

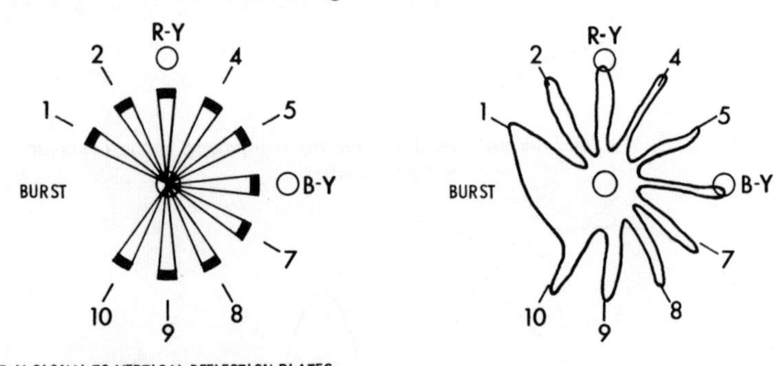

(A) *Theoretical daisy pattern.* (B) *Typical daisy pattern.*

Fig. 11-10. Vectorscope patterns.

If the vectorscope waveform is distorted, or if all the petals cannot be obtained, then trouble exists in the color demodulation section. This presumes, of course, that the tint control (or hue control as it is sometimes called) is properly adjusted.

Adjusting the color control of the receiver varies the amplitude of the color signal, and this, in turn, should change the size of the petals on the daisy. However, it should *not* change their angles. Varying the tint (or hue) control should rotate the petals, but should not change their amplitude. Thus, the daisy pattern is a good method of checking to see if these controls operate independently and correctly.

With a little practice you can align the color demodulator (but NOT the bandpass amplifier) by using the vectorscope display rather than aligning according to color bars as is the normal procedure.

Use of the Curve Tracer

A *curve tracer,* like the one shown in Fig. 11-11, is a valuable instrument for servicing transistorized equipment. With a curve tracer, it is possible to evaluate the condition of a transistor without removing it from the circuit. This is a strong advantage where removing the transistor from the circuit requires unsoldering.

Fig. 11-12 shows a simplified curve-tracer circuit, and characteristic curves obtained by this circuit. As shown in the circuit for the curve tracer (Fig. 11-12A), it employs a staircase generator and a full-wave bridge rectifier to obtain a family of curves for the transistor. Wideband dc oscilloscopes should be used in conjunction with the curve tracer to obtain the characteristic curves. The display shows the collector voltage versus the collector current. Since it is necessary to display the current on the oscilloscope, a small dropping resistor (100 ohms) is used to convert the current variations to voltage variations which can be displayed on the oscilloscope.

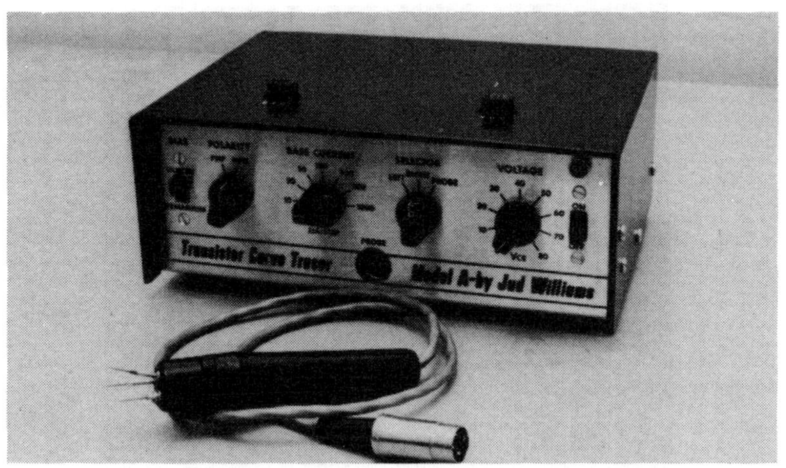

Fig. 11-11. A curve tracer used for troubleshooting transistor equipment.

Fig. 11-12B shows an ideal characteristic curve which is obtained by testing a transistor out of the circuit. However, it is not necessary to remove the transistor from the circuit in order to determine if it is good. Of course, the circuitry into which the transistor is connected may cause some distortion of the curve family, but it should be distinguishable as a characteristic curve. Fig. 11-13 shows a display that was obtained with a bad transistor. Here you see that the characteristic curve is no longer discernible. The use of a curve

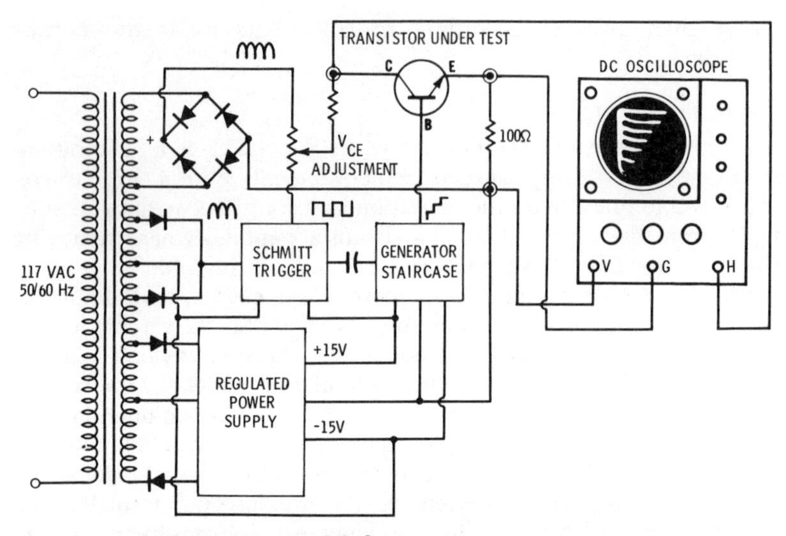

(A) Simplified curve tracer circuit.

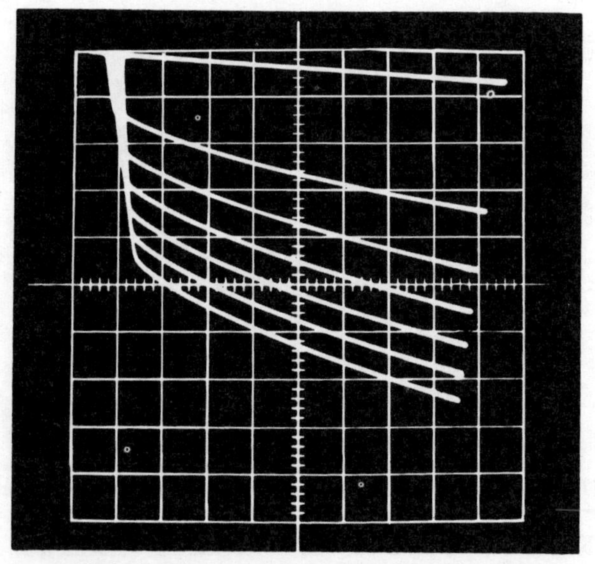

(B) Family of curves for a good transistor.

Fig. 11-12. Curve tracer circuit and an ideal pattern.

tracer as a test instrument requires a bit of practice, and it is necessary for you to learn to interpret the scope traces. However, the time that you will save by not having to connect and disconnect the transistors is well worth the effort.

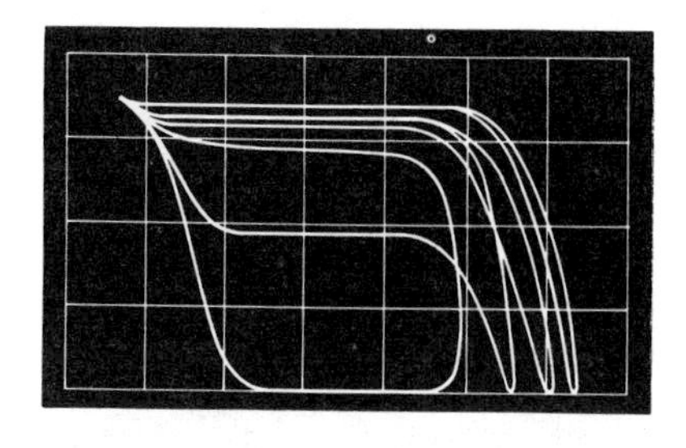

Fig. 11-13. An example of a curve obtained with a defective transistor.

Checking Inductances With a Ringing Test

There are ohmmeters available for checking resistance values, and capacitor checkers for determining if a capacitor is leaky, open, or within its capacitance rating. However, equipment for checking coils is not readily available. Fig. 11-14 shows how to use your oscilloscope to check a deflection coil, flyback, or other inductor in the receiver. This method of testing is called the *ringing test.* A pulse is generated across the coil at the start of the oscilloscope sweep. The best place to get this pulse is from the sawtooth output of the oscilloscope because this will mean that the damped wave produced will be automatically synchronized on the scope. However, a pulse from another source can also be used. (A square wave is an example of another source, and the square wave can be synchronized to the oscilloscope through the external sync terminals.)

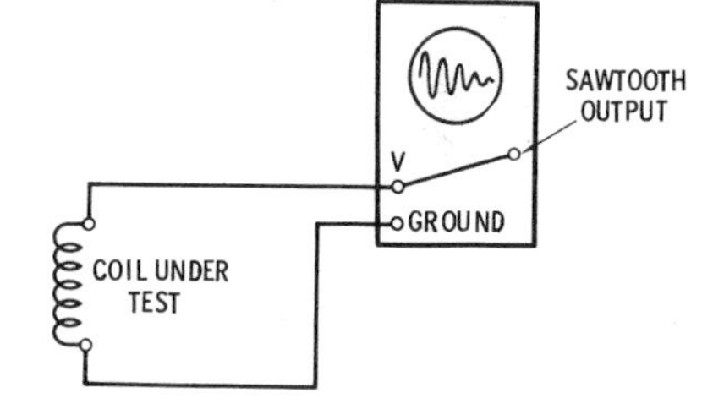

Fig. 11-14. Testing a coil with a ringing pattern.

Fig. 11-15 shows two curves obtained by the ringing test. The theory of the ringing test is that a pulse will cause the coil to produce oscillations such as the ones shown in Fig. 11-15A. Damped oscillations indicate ringing when the coil is pulsed. If the coil has shorted turns, or is defective in some other way, it will NOT produce the characteristic damped wave. Fig. 11-15B shows an example of the type of display obtained with a coil that has shorted turns.

As in the case of the curve tracer, it is necessary to practice this technique in order to be able to interpret properly the waveforms produced.

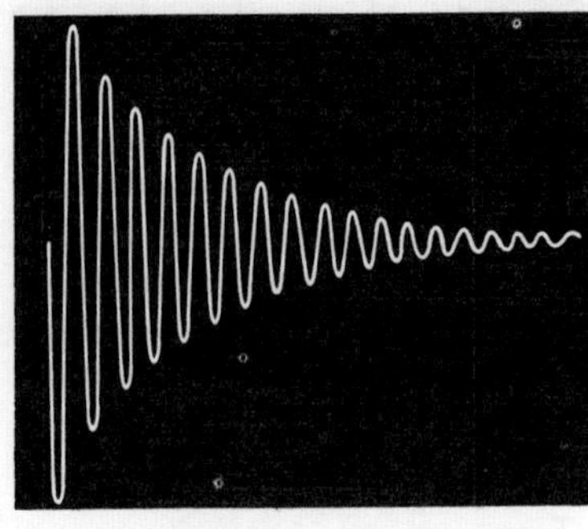

(A) *A good coil.*

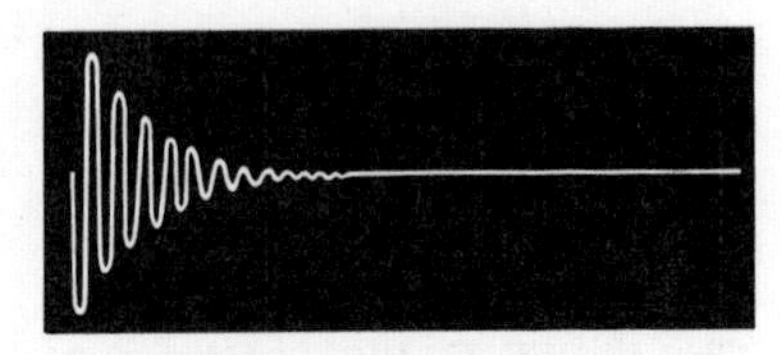

(B) *A shorted coil.*

Fig. 11-15. Typical ringing patterns for coils.

Square-Wave Testing

According to the Fourier Theory, any repetitive waveform can be reproduced by combining sine waves in the proper phases and amplitudes. A square wave is repetitive, and it can be produced by combining a fundamental sine wave with an infinite number of odd harmonics in the proper phase and amplitude. A very good square wave can be produced by using only the first six to ten odd harmonics.

Since the square wave can be considered to be a combination of frequencies, it can be used to test the response of an amplifier. This test is usually used as a *qualitative* rather than a *quantitative* test. In other words, it is possible to use the square wave test to tell if an amplifier has a poor high-frequency response or a poor low-frequency response, but it is not possible to accurately determine the actual cutoff frequencies from the shapes of the curves.

Chart 11-1 shows some typical patterns obtained by feeding a square wave through a video amplifier and observing the output on an oscilloscope. Two different square waves are used for the test. One is a low-frequency square wave that produces a pattern which makes it easy to determine if the low-frequency response of the amplifier is satisfactory. The horizontal line on the square wave represents periods when the amplifier must hold a d-c value for a relatively long time. If the low-frequency response of the amplifier is poor, it will not be able to hold this dc value and the result is a tilting of the horizontal lines.

The vertical lines of the square wave represent a very rapid change in voltage from one value to another. In order for an amplifier to pass this rapid change in voltage, it must have a good high-frequency response. Rounding or distortion of the vertical lines means that the amplifier cannot change values rapidly enough to accommodate the voltage change, and therefore, the high-frequency response is questionable. Shown in Chart 11-1 are some typical problems with amplifiers having poor high-frequency response.

One technique used with a square-wave generator is to apply the square-wave signal to the input of the first video amplifier and observe the waveshapes shown on the picture tube. Fig. 11-16 shows some representative patterns. Fig. 11-16A is the pattern obtained when a 60-hertz square wave is applied to the video-amplifier chain.

Chart 11-1. Square-wave response patterns.

INPUT TO VIDEO AMPLIFIER	SHAPE OF OUTPUT WAVE SEEN ON CRT SCREEN	INTERPRETATION
 60Hz SQUARE WAVE		GOOD LOW-FREQUENCY RESPONSE AND NEGLIGIBLE PHASE SHIFT
		LEADING LOW-FREQUENCY PHASE SHIFT AND LOW-FREQUENCY ATTENUATION
		LAGGING LOW-FREQUENCY PHASE SHIFT
 25 kHz SQUARE WAVE		GOOD HIGH-FREQUENCY AND TRANSIENT RESPONSE
		POOR HIGH-FREQUENCY RESPONSE
	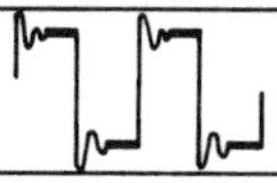	EXCESSIVE HIGH-FREQUENCY RESPONSE (OSCILLATION) AND NONLINEAR TIME DELAY
		EXCESSIVE OR INSUFFICIENT MIDFREQUENCY RESPONSE AND NONLINEAR TIME DELAY

If the response of the video amplifier is good, there should be even shading for the dark and light areas produced. Uneven shading means that the amplifier cannot hold the required dc value during the trace time for that shade. Fig. 11-16B shows the result of feeding a square wave through the video amplifier section that has poor low-frequency response.

If a 15,750-hertz square wave is applied through the video amplifiers, a signal like that of Fig. 11-16C will be obtained. It is possible that this pattern may be reversed so that the black is on the right side and the white is on the left side. (This is also true of Fig. 11-16A.) The reason that pattern reversal may occur is that the number of video stages between the point where the square wave is injected and the cathode ray tube determines the voltage polarity needed for

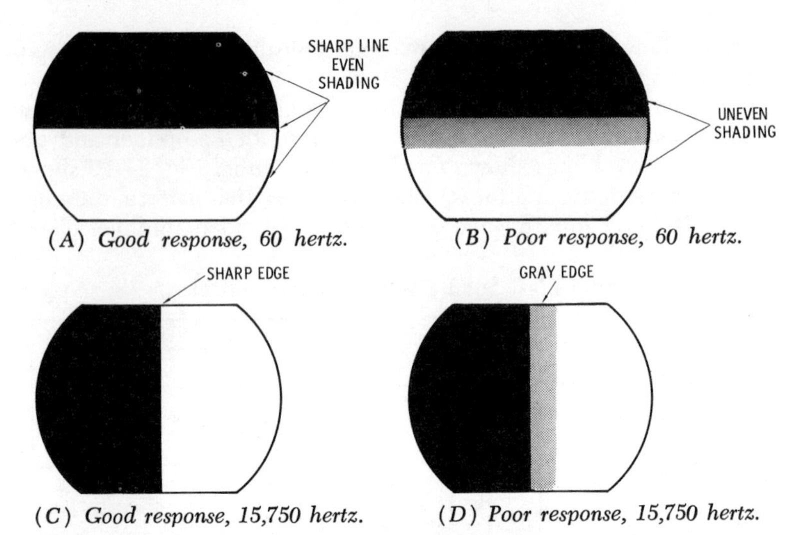

(*A*) *Good response, 60 hertz.*

(*B*) *Poor response, 60 hertz.*

(*C*) *Good response, 15,750 hertz.*

(*D*) *Poor response, 15,750 hertz.*

Fig. 11-16. Using the picture tube as an indicator for the square-wave test.

picture-tube cutoff. Another determining factor is whether the signal is fed to the grid or to the cathode of the picture tube.

Returning to the pattern of Fig. 11-16C, if the frequency response of the receiver is satisfactory, the pattern will pass from the black to the white (or from white to black) very rapidly as indicated by a sharp edge on the pattern. If the response is poor, then the pattern will show a fuzzy edge as in Fig. 11-16D. In the case of excessive high-frequency response, there will be ghost-like lines beside the vertical line.

By increasing the frequency of the square wave many horizontal (or vertical) stripes will be obtained. These lines can be used for adjusting the linearity and size. To produce horizontal lines, the square-wave frequency must be less than 15,750 Hz. To produce vertical lines, the square-wave frequency must be greater than 15,750 Hz.

The square-wave method of testing an amplifier is an interpretative test—that is, the success of using this test depends on the ability of the technician to interpret the results. It is necessary to practice the technique so that you can get an idea of what is expected in terms of results.

PROGRAMMED QUESTIONS AND ANSWERS

Starting with question number 1, select the answer that you feel is correct. If you feel that (A) is correct, proceed to block number 17 as directed. If you feel that (B) is correct, proceed to block number 9 as directed. If you feel that more than one answer is correct, choose the one that you think is the *most* correct.

. . . .

1 In order to employ a post marker, which test instrument would you use?
(A) A marker post. (Go to block number 17.)
(B) A marker adder. (Go to block number 9.)

. . . .

2 Your answer is wrong. You may not get an ideal curve like the one shown by the manufacturer, but you should get a reasonable looking curve with the transistor in the circuit. Go to block number 7.

. . . .

3 Your answer is wrong. The horizontal lines on the square wave are related to the low-frequency response. Go to block number 23.

. . . .

4 The initials "VSM" stand for Video Sweep Modulator. This is a method of obtaining a bandpass-amplifier response curve with an rf sweep generator and marker generator.
Here is your next question . . .
A sine-wave signal is applied to an audio amplifier with a resistive load. The input and output signals are used to obtain a Lissajous pattern. For this particular amplifier, the pattern obtained is an ellipse. Which of the following is true?
(A) There is phase shift distortion in the amplifier. (Go to block number 15.)
(B) This is a normal pattern. (Go to block number 21.)

. . . .

5 Your answer is wrong. It is true that excessive high-frequency response will cause oscillations on the square-wave pattern, and that this is sometimes called *ringing*. However, the ringing test is used for coils. Go to block number 11.

. . . .

6 Your answer is wrong. Study the illustrations in Fig. 11-6, then go to block number 24.

. . . .

7 Usually, the transistor can be tested while it is connected into the circuit.

Although the curve tracer method of troubleshooting has been used for a number of years in some electronics industries, it is a relatively recent method in troubleshooting home-entertainment equipment.

Here is the next question . . .

A color bandpass amplifier should have a frequency response about:

(A) 1.2-MHz wide. (Go to block number 13.)

(B) 4.2-MHz wide. (Go to block number 19.)

. . . .

8 Your answer is wrong. The tuner should be able to produce the required response on any channel for which the receiver is being used. Go to block number 14.

. . . .

9 The disadvantage of the post marker system is that it requires a marker adder, and this is an additional investment in test equipment. However, in terms of obtaining an accurate response curve, this method is preferred.

Here is the next question . . .

To obtain a bandpass-amplifier response curve, you might use a method called:

(A) VSM. (Go to block number 4.)

(B) VOM. (Go to block number 10.)

. . . .

10 Your answer is wrong. A VOM is a volt-ohm-milliammeter. This is a test *instrument,* not a test method. Go to block number 4.

. . . .

11 The ringing test can be used for determining if there is a short in the deflection coil.

Here is the next question . . .

A 15,750-Hz square-wave signal is fed through the video amplifier stages, and the pattern is observed on the screen. This pattern is used for checking:

(A) high-frequency response. (Go to block number 16.)

(B) low-frequency response. (Go to block number 22.)

. . . .

12 Your answer is wrong. The I and R − Y signals are for two different receiver systems. Go to block number 18.

. . . .

13 As shown in Fig. 11-7, a typical bandpass-amplifier response curve is only about 1.2-MHz wide.

Here is the next question . . .

A notch will be placed in a response curve when:

(A) an absorption marker is used. (Go to block number 24.)

(B) a crystal oscillator is used for a marker. (Go to block number 6.)

. . . .

14 A steady dc voltage may be required for biasing the rf amplifier in the tuner. In some tuners the bias voltage may not be needed. This is true for receivers in which only the i-f amplifier is agc biased.

Here is the next question . . .

To obtain a circle Lissajous pattern, two sine waves are applied to an oscilloscope. These waveforms must be:

(A) in phase. (Go to block number 20.)

(B) 90° out of phase. (Go to block number 25.)

. . . .

15 The ellipse indicates that the phase shift is not 180° or 0°, and there should be a further investigation of the amplifier.

Here is the next question . . .

When using a curve tracer for troubleshooting,

(A) a transistor under test must be removed from the circuit. (Go to block number 2.)

(B) a transistor can usually be tested while it is still connected into its circuit. (Go to block number 7.)

. . . .

16 With a 15,750-Hz square-wave signal, a black-and-white pattern with a vertical division will be observed. If the receiver has a satisfactory high-frequency response, a sharp division should exist between the black-and-white areas.

Here is the next question . . .

The sync pulses at the output of the video detector must be the most positive part of the signal. (True or False) (Go to block number 26.)

. . . .

17 Your answer is wrong. There is probably no such thing as a "marker post" unless it is a terminal on the sweep generator. Go to block number 9.

. . . .

18 Your answer is correct. The ten petals are generated by feeding the color-bar generator signal to the receiver in the normal manner, and then comparing the R − Y and B − Y signals in a Lissajous pattern.

Here is the next question . . .

In a square-wave test, poor high-frequency response is indicated on the oscilloscope by:

(A) distortion of the vertical lines of the square wave. (Go to block number 23.)

(B) distortion of the horizontal lines of the square wave. (Go to block number 3.)

. . . .

19 Your answer is wrong. A typical response curve for a bandpass amplifier is shown in Fig. 11-7. Go to block number 13.

. . . .

20 Your answer is wrong. When the waves are in phase a straight line is obtained. Go to block number 25.

. . . .

21 Your answer is wrong. The phase shift in an amplifier should be 180° (for common-cathode and common-emitter configurations) or 0° (for common-grid or plate and common-base or common-collector configurations). All of these cases should produce a straight line. Go to block number 15.

. . . .

22 Your answer is wrong. Study Fig. 11-16, then go to block number 16.

. . . .

23 The vertical lines correspond to a very rapid change in voltage. An amplifier must have a good high-frequency response in order to reproduce these lines.

Here is the next question . . .

To obtain the correct tuner response curve:

(A) always run the test with the tuner set to the middle of the vhf response curve. (Go to block number 8.)

(B) it is usually necessary to supply a steady dc voltage to the agc terminal. (Go to block number 14.)

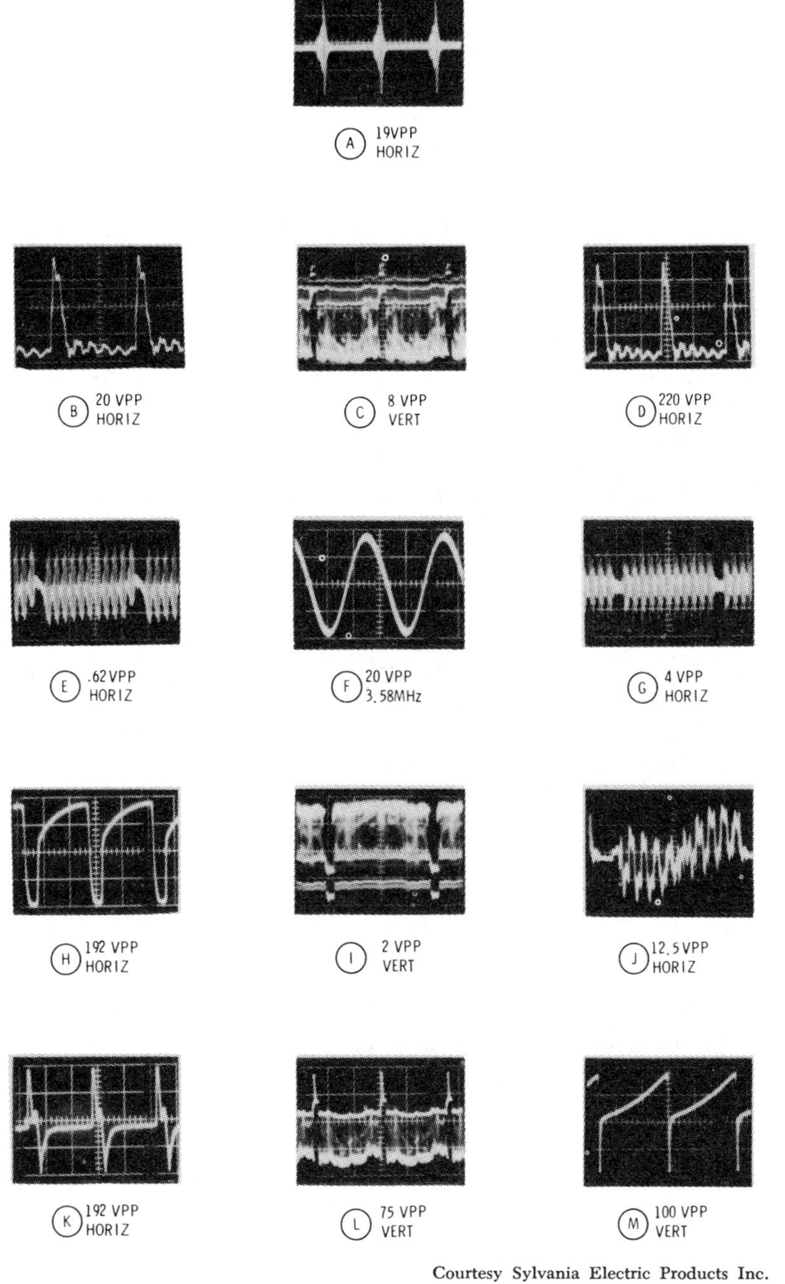

Courtesy Sylvania Electric Products Inc.

Fig. 11-17. Waveforms associated with receiver schematic shown in Fig. 11-18.

. . . .

24 Since the absorption marker will not pass the frequency to which it is set, a notch is produced on the response curve.

Here is the next question . . .

The vectorscope pattern is:

(A) obtained when the I and R-Y signals are compared in a Lissajous pattern. (Go to block number 12.)

(B) a Lissajous pattern that shows ten petals which are due to the ten color bars generated by a color-bar generator. (Go to block number 18.)

. . . .

25 When the two waves are 90° out of phase, a circle will be obtained.

Here is the next question . . .

The ringing test can be used to determine:

(A) if the video amplifier will pass all frequencies in the video range. (Go to block number 5.)

(B) if the deflection coil has shorted turns. (Go to block number 11.)

. . . .

26 The correct answer is false. The polarity of the signal is dependent upon the number of video amplifier stages used, and it also is determined by whether the signal is fed to the cathode or control grid of the picture tube.

You have now completed the programmed questions and answers.

. . . .

PRACTICE TEST

1. If the video amplifier stage in a receiver has a poor low-frequency response,

(a) the receiver will be unable to hold vertical sync.
(b) the receiver will be unable to hold horizontal sync.
(c) the vertical blanking bar will not have a constant shade of gray from start to finish.
(d) the detector diode is probably leaky.

2. If the video signal is applied to the cathode of the picture tube, the sync pulses:

(a) will be the most positive part of the signal.
(b) will be the most negative part of the signal.

3. An advantage to using a synchroscope for viewing the color burst is that:

(a) it has a higher frequency response.
(b) it makes it possible to lock the synchroscope onto the burst.
(c) the synchroscope shows the burst in a different color.
(d) the synchroscope shows the color lobes in a "daisy pattern."

4. The color burst is located:

(a) on the back porch of the vertical sync pulse.
(b) on the front porch of the vertical sync pulse.
(c) on the back porch of the horizontal sync pulse.
(d) on the front porch of the horizontal sync pulse.

5. The horizontal deflection coils are located:

(a) at the top and bottom of the yoke.
(b) at the right and left sides of the yoke.
(c) only at the top side of the yoke.
(d) only at the left side of the yoke.

6. Markers are placed on the response curve during sweep alignment in order to:

(a) blank the scope during retrace.
(b) identify certain frequencies on the response curve.
(c) eliminate the back voltage generated in the yoke during retrace.
(d) identify certain amplitudes on the response curves.

7. A square wave is applied to an oscilloscope to check the high-frequency response of its vertical amplifier. Poor high-frequency response is indicated:

(a) by rounding off of the leading and trailing edges.
(b) when the top and bottom of the square wave are slanting toward the center line.
(c) if there is a loss of amplitude for the square wave.
(d) when the frequency of the square wave is decreased by the scope.

8. The voltage at the output of the video detector in a television receiver should be measured with:

(a) a VTVM to determine the rms value of the signal.
(b) a VOM to determine the rms value of the signal.
(c) a milliammeter to determine the rms value of the signal.
(d) an oscilloscope to determine the peak-to-peak voltage of the video signal.

9. Which of the following represents the most logical frequency range for a color bandpass amplifier?

(a) 3.0 – 4.1 MHz.
(b) 1.0 – 4.5 MHz.
(c) 0 – 1.2 MHz.
(d) 4.1 – 4.5 MHz.

10. In the composite video signal, the sync pulse:

(a) has about the same amplitude as the blanking pedestal.
(b) has about the same amplitude as the video signal.
(c) has about the same amplitude as the color burst.
(d) is generated with eight cycles of carrier.

11. A luminosity response curve of the eye would show maximum response at:

(a) the blue wavelength.
(b) the red wavelength.
(c) the green-yellow wavelength.
(d) the ultraviolet end of the spectrum.

12. For the composite television signal, which of the following has the greatest bandwidth?

(a) The I signal.
(b) The Q signal.
(c) The audio signal.
(d) The luminance signal.

13. The gradual slope on the receiver response curve, with the picture carrier in the center of the slope, is needed to:

(a) prevent the picture carrier from heterodyning with the signal.
(b) compensate for the fact that the television signal is transmitted with a vestigial sideband.
(c) give the color signals greater amplification than the luminance signal.
(d) prevent adjacent sideband interference.

14. A Lissajous pattern is formed on an oscilloscope by:

(a) feeding a sine-wave voltage to the vertical deflection plates and a sawtooth wave to the horizontal deflection plates.
(b) feeding a sine-wave voltage to the horizontal deflection plates and a sawtooth wave to the vertical deflection plates.
(c) feeding a sine-wave voltage to both the vertical and the horizontal deflection plates.
(d) feeding a sawtooth current to both the vertical and the horizontal deflection plates.

15. A circle is obtained on the scope when

(a) two in-phase sine waves are fed to the horizontal and the vertical deflection plates.
(b) two sine waves that are 90° out of phase are fed to the vertical and horizontal deflection plates.
(c) two sawtooth waves, in phase, are fed to the vertical and the horizontal deflection plates.
(d) a sine wave and a sawtooth wave are compared on the scope screen.

16. To reproduce white on the screen of the color picture tube:

(a) all three guns must be conducting.
(b) only the green and blue guns are conducting.
(c) the color burst signal must be at its maximum amplitude.
(d) the voltages of the color difference signals in the composite signal must be maximum.

17. Which is greater?

(a) The rms value of a sine-wave voltage.
(b) The average value of a sine-wave voltage.

18. The peak-to-peak value of a sine wave is measured with an oscilloscope and found to be 15 volts. What is the rms value of this voltage?

(a) 10.6 V.
(b) 9.54 V.
(c) 7.2 V.
(d) 5.3 V.

19. A true square wave could be obtained by:

(a) combining two sawtooth waves.
(b) combining a fundamental-frequency sine wave with an infinite number of odd harmonic sine waves, each having the proper amplitude and phase.

(c) combining a fundamental-frequency sine wave with an infinite number of even harmonic sine waves, each having the proper amplitude and phase.
(d) combining a fundamental-frequency sine wave with both even and odd harmonic frequencies, each having the proper phase and amplitude.

20. The waveform of the voltage delivered to the deflection yoke is not a sawtooth. Instead, it is a sawtooth sitting on a square pedestal. The reason for using this type of waveform is:

(a) to overcome inductive kickback in the yoke.
(b) to increase the rms value of the waveform.
(c) to produce the necessary ringing in the yoke during retrace.
(d) to decrease the amount of heating that will take place in the yoke.

21. Oscillation would occur in the horizontal deflection coil:

(a) if the sync pulse were not present at the horizontal oscillator.
(b) if the flyback transformer were not properly grounded.
(c) if a thermistor was not placed in the yoke.
(d) if a damper tube was not used.

22. Pulses for vertical synchronization:

(a) come directly from the detector.
(b) come directly from the sync separator.
(c) come from an integrator.
(d) come from the flyback transformer.

23. Pairing of the interlacing is evident on the test pattern by checking the:

(a) vertical lines.
(b) horizontal lines.
(c) diagonal lines.
(d) edge of light and dark areas for ringing.

24. The petals of the "daisy" on the vector scope are rotating. Which of the following is true?

(a) This is what it is supposed to do.
(b) The sync separator is not working properly.
(c) The 3.58-MHz oscillator is not synchronized.
(d) This is impossible.

25. To align the chroma bandpass amplifiers you should use:

(a) a vectorscope and a color-bar generator.
(b) a synchroscope and a curve tracer.
(c) a sweep generator and an oscilloscope.
(d) an accurate rf generator and a VTVM.

26. If a few turns in the yoke are shorted, the best check is to:

(a) use a ringing test.
(b) use a megger.
(c) use an ohmmeter.
(d) measure the receiver B+ voltage.

27. If you do not have a dual-trace scope, you can look at two waveforms simultaneously on a regular oscilloscope with the aid of:

(a) a mixer.
(b) a nonlinear resistor.
(c) a detector probe.
(d) an electronic switch.

28. The sine-wave input signal to a Class-A audio amplifier with a pure resistance load is fed to the horizontal amplifier of an oscilloscope. The output signal is fed to the vertical amplifier. The gain adjustments are set to produce a two-inch trace for each input. If there is no distortion in the amplifiers, the trace will be:

(a) an ellipse.
(b) a straight line.
(c) a circle.
(d) a figure eight pattern.

29. A pure sine-wave voltage is fed to a differentiating circuit. The output signal will be:

(a) a square wave.
(b) a sawtooth.
(c) a pulse.
(d) a sine waveform.

30. The ripple frequency of a full-wave rectifier circuit in a 400-Hz power system would be:

(a) 400 Hz.
(b) 800 Hz.
(c) 60 Hz.
(d) 120 Hz.

31. An oscilloscope is calibrated for a ten volt-per-inch deflection. If a ten-volt (rms) sine-wave voltage is fed to the vertical terminals of the scope, the deflection should be:

(a) about 1.4 inches.
(b) about 1.11 inches.
(c) about 2.8 inches.
(d) exactly 10 inches.

32. When you are sweep aligning the i-f section of a receiver,

(a) always start with the audio section.
(b) never change the adjustments of the traps.
(c) always disconnect the yoke to prevent interference from the sweep sections.
(d) start at the detector input stage and work toward the tuner.

33. To check the high-frequency response of the video-amplifier stage, use:

(a) a sweep generator with markers.
(b) an rf generator and a VOM.
(c) a color generator.
(d) a square-wave generator.

34. One difference between a synchroscope and an oscilloscope is that:

(a) the trace on a synchroscope is vertical, and the trace on the oscilloscope is horizontal.
(b) when there is no signal applied, the scope has a trace and the synchroscope does not.
(c) the synchroscope cannot be used for voltage measurements.
(d) there is no *DC INPUT* provision on the oscilloscope, but there is on the synchroscope.

For questions numbered 35 through 50 inclusive, you are to select your answer after referring to the waveforms shown in Fig. 11-17, and the receiver schematic of Fig. 11-18 (refer to foldout in back of book). The overall schematic of the Sylvania television receiver from which the waveforms were taken is shown in Fig. 11-18. The terms *VERT.* and *HORIZ.* refer to the oscilloscope sweep-frequency setting used to obtain the waveform. An important clue to the identity of the waveform may be the peak-to-peak voltage which is also given with each waveform.

As an additional exercise, the overall schematic should be studied carefully in regard to the typical problems identified with arrows. In a CET or licensing exam, you may be asked which type of problem is associated with a certain symptom.

35. In regard to the blanker circuit (Q604) which of the following is true?

(a) The transistor is biased for Class-A operation.
(b) The way the transistor is biased, it is normally cut off in the absence of an input signal.

36. The amount of forward bias on the first chroma amplifier (Q610 in Fig. 11-18) is:

(a) 2.5 volts.
(b) 1.8 volts.
(c) 0.7 volts.
(d) 4.3 volts.

37. The amount of bias on the second video amplifier: (Q208 in Fig. 11-18) is:

(a) +10.3 volts.
(b) −0.7 volts.
(c) 16.8 volts.
(d) 6.5 volts.

38. From the choices in Fig. 11-17, select the waveform that will appear at the output signal from the 3.58-MHz oscillator (Point 32 on the schematic).

39. From the choices in Fig. 11-17, select the waveform that will appear at the plate of the vertical oscillator (Point 6 on the schematic).

40. From the choices in Fig. 11-17, select the waveform that will appear at the input of the first video amplifier (Point 1 on the schematic).

41. From the choices in Fig. 11-17, select the waveform that will appear at the input to the sync separator (Point 3 on the schematic).

42. From the choices in Fig. 11-17, select the waveform that will appear at the input of the afc. (Point 11 on the schematic). This signal comes from the flyback transformer T400.

43. From the choices in Fig. 11-17, select the waveform that will appear at the input to the horizontal output (Point 14 on the schematic).

44. From the choices in Fig. 11-17, select the waveform that will appear at the plate of the video output (Point 16 on the schematic).

45. From the choices in Fig. 11-17, select the waveform that will appear at output signal from the first chroma amplifier. (Point 17 on the schematic).

46. From the choices in Fig. 11-17, select the waveform that will appear at the input to the R − Y amplifier (Point 31 on the schematic).

47. From the choices in Fig. 11-17, select the waveform that will appear at the output of the burst amplifier (Point 27 on the schematic).

48. From the choices in Fig. 11-17, select the waveform that will appear at the plate of the horizontal output stage (Point 15 on the schematic).

49. From the choices in Fig. 11-17, select the waveform that will appear at the input to the blanker (Point 25 on the schematic).

50. From the choices in Fig. 11-17, select the waveform that will appear at the output signal from the second chroma amplifier (Point 18 on the schematic).

12

Practice Test

KEYED STUDY ASSIGNMENT

There is no keyed study assignment for this chapter.

Chapter 12 is a representative exam which covers the material you will be expected to know in CET and licensing exams. These questions are not, of course, the actual questions that you will encounter on the exam.

You should grade each section of this exam separately. If more than eight questions are missed in any section, you should review that section and the reading assignments associated with it before attempting to sit for either the CET exam or a licensing exam.

PART I—THE TELEVISION SIGNAL

1. A time interval that is used as a standard unit of time when measuring other time intervals of the composite television signal is:

(a) 10 microseconds.
(b) 10.7 microseconds.
(c) 27.5 microseconds.
(d) 63.5 microseconds.

2. The part of the composite video signal that is usually used as reference for the receiver agc is:

(a) the blanking pedestal.
(b) the video signal.
(c) the white reference level of the video signal.
(d) the sync pulses.

3. A test signal that is transmitted by the television station is the:

(a) STI.
(b) SCA.
(c) VOTS.
(d) VITS.

4. In order to make it possible to interleave the color information into the existing monochrome signal:

(a) the horizontal sweep frequency was lowered slightly.
(b) the horizontal sweep frequency was raised slightly.

5. The exact value of vertical sweep for a color signal is:

(a) 60 hertz.
(b) 59.94 hertz.
(c) 60.15 hertz.

6. The time for one horizontal line is:

(a) $\frac{1}{15{,}750}$ microseconds.
(b) $\frac{1}{15{,}750}$ milliseconds.
(c) $\frac{1}{15{,}750}$ seconds.
(d) $\frac{1}{15{,}750} \times 10^{-3}$ seconds.

7. The frame frequency for the television signal is:

(a) 60 hertz.
(b) 15,750 hertz.
(c) 30 hertz.
(d) 30.14 hertz.

8. As the signal is transmitted, the sound frequency is:

(a) 4.5 MHz above the video carrier frequency.
(b) 4.5 MHz below the video carrier frequency.
(c) equal to the video carrier frequency.
(d) 2.3 MHz below the color subcarrier.

9. In the receiver i-f stages, the sound frequency is:

(a) 4.5 MHz above the video carrier frequency.
(b) 4.5 MHz below the video carrier frequency.
(c) equal to the video carrier frequency.

10. On the i-f response curve, the video frequency is at one-half the maximum amplitude. The response curve is purposely made this way because:

(a) too much carrier amplitude could overdrive the receiver.
(b) the television signal is transmitted single sideband and this shape of response curve is necessary in order to produce a flat response to video frequencies.
(c) the television signal is transmitted vestigial sideband and this shape of response curve is necessary in order to produce a flat response to video frequencies.
(d) it reduces the possibility of heterodyning between the video and sound signals.

11. Which of the following statements is not correct?

(a) The audio frequency being transmitted by an fm station is determined by the number of times per second that the carrier crosses the center frequency.
(b) To demodulate the left channel of the stereo fm signal, the (L + R) and (L − R) signals are combined by subtraction.
(c) The SCA signal is used for background music.
(d) The L − R signal is sent as an a-m signal with two sidebands, but without a carrier.

12. The color burst signal is transmitted:

(a) on the front porch of the horizontal blanking pedestal.
(b) on the back porch of the horizontal blanking pedestal.
(c) on the front porch of the vertical blanking pedestal.
(d) on the back porch of the vertical blanking pedestal.

13. The time required for a complete television frame is _____ microseconds.

14. The color burst frequency is:

(a) 60 kHz.
(b) 3.85 kHz.
(c) 3.58 MHz.
(d) 5.83 MHz.

15. Which of the following receiver customer controls, if incorrectly adjusted, can result in a loss of color?

(a) The fine tuning control.
(b) The contrast control.
(c) The brightness control.
(d) The volume control.

16. The color burst signal is used for:

(a) keeping the tuner local oscillator in step with the transmitted video carrier.
(b) keeping the receiver color oscillator in step with the color oscillator at the transmitter.
(c) supplying the receiver with the I and Q color signals.

17. The sync pulses represent _____% of the total composite television signal amplitude.

18. The blanking pedestals represent _____% of the total composite television signal amplitude.

19. The minimum amplitude of the composite video signal is _____% of the maximum amplitude.

20. The type of transmitted signal with the NTSC signal is:

(a) positive picture transmission.
(b) negative picture transmission.

21. The sound i-f frequency in an intercarrier receiver is:

(a) 455 kHz.
(b) 4.5 MHz.
(c) 10.7 MHz.
(d) 41 MHz.

22. The primary colors in the NTSC system are ____________________.

23. The reason for not allowing the carrier amplitude to drop to zero during transmission of white in a monochrome picture is:

(a) the subcarrier would be lost.
(b) intercarrier receivers require at least a 10-percent video carrier in order to be able to demodulate the sound.
(c) the agc voltage in receivers with separate sound becomes excessive when the carrier level drops below 10 percent maximum.

24. Fig. 12-1 shows a line of video signal. Which of the following shows the appearance of the tones produced by this line?

(a) Choice A.
(b) Choice B.
(c) Choice C.
(d) Choice D.

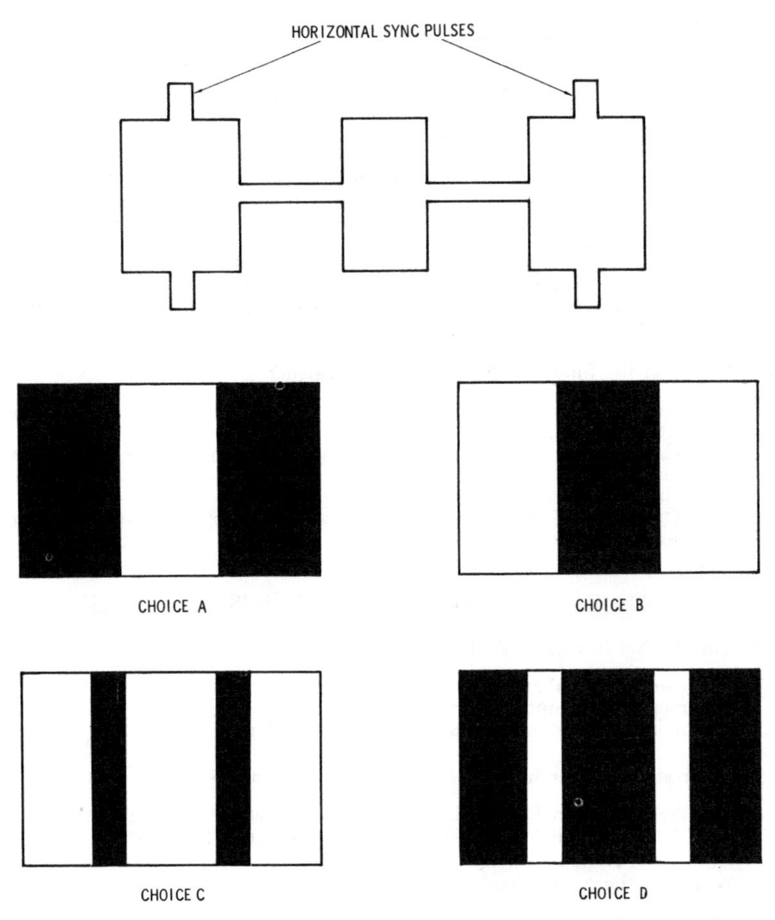

Fig. 12-1. Choose the correct pattern for the video waveform shown at the top.

25. Which of the following types of television transmission is more likely to produce better synchronization?

(a) Negative modulation. (b) Positive modulation.

26. A certain receiver employs keyed agc. A scene with a dancer dressed in white against a black background is being televised. When several more dancers, similarly attired, enter the scence:

(a) the receiver agc voltage reduces the receiver gain.
(b) the receiver agc voltage remains the same.
(c) the receiver agc voltage becomes more negative.

27. A jagged diagonal line on the test pattern displayed by a receiver indicates:

(a) poor low-frequency response.
(b) poor high-frequency response.
(c) low-frequency phase shift.
(d) pairing of the interlacing.

28. The luminance signal is also called:

(a) the I signal.
(b) the Q signal.
(c) the Y signal.
(d) the SCA.

29. Which of the following is not a correct statement?

(a) The Q signal is limited to 0.5 MHz and both sideband are transmitted.
(b) For maximum detail, the luminance signal is sent with wideband transmission.
(c) The I signal represents colors of orange and cyan, and one of the sidebands is partially eliminated at the transmitter.
(d) The makeup of the luminance signal is $E_Y = 0.59\,E_R + 0.30E_G + 0.11E_B$.

30. Which of the following is not a true statement?

(a) Very small areas of the color picture are transmitted in monochrome rather than in color.
(b) The I and Q signals are combined vectorially to produce the chrominance signal.
(c) The color burst is only for the Q signal.
(d) The reference color burst signal is transmitted by amplitude modulating the composite television signal.

PART II—ANTENNAS AND TRANSMISSION LINES

1. The decibel is a method of:

(a) comparing voltages, currents, or powers.
(b) measuring sound power.
(c) measuring antenna voltage, current, or power.

2. A certain amplifier has a gain of −3dB. This means that:

(a) the input signal power is greater than the output signal power.
(b) the input signal power is less than the output signal power.

3. Which of these equations is not correct?

(a) $dB = 20 \log_{10} \frac{P_2}{P_1}$.
(b) $dB = 20 \log_{10} \frac{E_2}{E_1}$.
(c) $dB = 20 \log_{10} \frac{I_2}{I_1}$.

4. In a certain amplifier, the output power is equal to the input power. The dB gain is:

(a) 0 dB.
(b) 2 dB.
(c) 4 dB.
(d) 1.0 dB.

5. In a certain transmission line, the power at the output end is 500 microwatts, and the signal at the input end is 2000 microwatts. Which of the following describes the dB gain of the line?

(a) 6 dB.
(b) −6 dB.
(c) 3 dB.
(d) −3 dB.

6. A certain transmission line has a loss of 9 dB. If the power at the output end is 800 microwatts, then the signal at the input end is:

(a) 100 microwatts.
(b) 6400 microwatts.
(c) 7200 microwatts.
(d) 80 microwatts.

7. The output voltage of a certain amplifier is 1.8 volts. If the gain of the amplifier is 6 dB, the input voltage must be:

(a) 9 volts.
(b) 0.9 volts.
(c) 90 millivolts.
(d) 900 microvolts.

8. The fm broadcast band is:

(a) between Channels 6 and 7.
(b) between Channels 5 and 6.
(c) between Channels 7 and 8.
(d) between the vhf and uhf bands.

9. The television signal is:

(a) horizontally polarized.
(b) vertically polarized.

10. Which of the following types of antennas is not normally used as a standard for calculating antenna gain?

(a) An isotropic.
(b) A Yagi.
(c) A simple dipole.

11. In a strong-signal area with no ghost signals, which of the following antennas would be best for receiving a color signal on Channel 4 and Channel 5?

(a) A Yagi.
(b) A log periodic.

12. The directors of an antenna are:

(a) between the driven element and the reflector.
(b) between the driven element and the transmitter.
(c) on the opposite side of the driven element from the transmitter.
(d) used for matching the antenna and transmission line impedances.

13. Which of the following types of transmission lines is most susceptible to losses when foreign materials are deposited on its surface?

(a) Flat twin lead.
(b) Foam-filled twin lead.
(c) Coaxial cable.
(d) Shielded twin lead.

14. Which of the following types of transmission line is used for MATV systems?

(a) Foam-filled twin lead.
(b) Coaxial cable.

15. Which of the following is an unbalanced transmission line system?

(a) Flat twin lead.
(b) Coaxial cable.

16. To match a balanced transmission line to an unbalanced transmission line, use a ____________.

17. Standing waves on a transmission line are caused by:

(a) excessive wind and an insufficient number of standoffs.
(b) noise signals.
(c) impedance mismatching.
(d) excessive signal from the antenna.

18. Two main sections of an MATV system are the ________ and the ________.

19. The standard of signal strength in an MATV system is ____ microvolts.

20. When a wave is vertically polarized, its ________ (electric or magnetic) field is perpendicular to the surface of the earth.

21. When two different stations in different areas are received simultaneously on the dial setting, the phenomenon is known as:

(a) sporadic video.
(b) dual-channel interference.
(c) intrachannel interference.
(d) co-channel interference.

22. The distance that an electromagnetic wave travels in one cycle is called the ________.

23. Which of the following elements of a Yagi antenna is shortest in length?

(a) A director.
(b) A driven element.
(c) A reflector.

24. For a half-wave dipole antenna:

(a) the voltage at the center is minimum.
(b) the current at the center is minimum.
(c) the correct length of a half-wave dipole does not depend on the proximity of other objects.

25. A simple dipole antenna is shown in Fig. 12-2.

(a) The arrows show the direction of signals that will produce maximum signal strength at the receiver.
(b) The arrows show the direction of signals that will produce minimum signal strength at the receiver.

DIPOLE

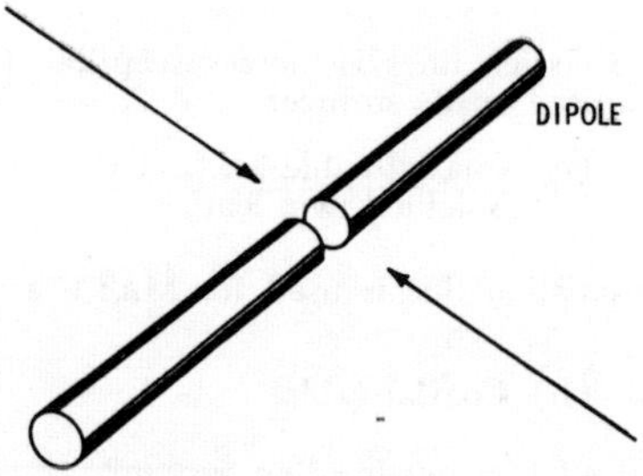

Fig. 12-2. A simple dipole antenna.

26. The impedance of a simple half-wave dipole at the center is:

(a) 50 ohms.
(b) 72 ohms.
(c) 150 ohms.
(d) 300 ohms.

27. The impedance of a folded dipole at the transmission line terminals is:

(a) 50 ohms.
(b) 72 ohms.
(c) 150 ohms.
(d) 300 ohms.

28. In comparing a simple dipole with a folded dipole:

(a) they both have the same impedance.
(b) they both have the same gain.
(c) they both have the same reception pattern.
(d) they are both color-coded green.

29. Stacked arrays do not provide:

(a) higher gain.
(b) improved directional characteristics.
(c) lower wind resistance.

30. Which of the following types of transmission lines will exhibit the maximum amount of loss from moisture that has accumulated on the surface?

(a) Flat-ribbon parallel line.
(b) Tubular parallel line.
(c) Encapsulated parallel line.
(d) Coaxial cable.

PART III—ELECTRONIC COMPONENTS

1. Which of the following capacitor ratings indicates that the capacitance will increase as the temperature of the capacitor increases?

(a) NPO.
(b) 220 pF.
(c) N750.
(d) P100.

2. When the fourth color band is gold on a carbon composition resistor, it means that the resistor:

(a) has a resistance value less than 10 ohms.
(b) has a resistance value less than 1.0 ohm.
(c) has a tolerance of 10 percent.
(d) has a tolerance of 5 percent.

3. In a television receiver, an integrator is:

(a) a high-pass RC filter.
(b) a low-pass RC filter.
(c) a circuit that removes the vertical sync pulse from the blanking pedestal.
(d) a circuit that combines the burst signal with the I and Q signals.

4. A resistor used for converting a microammeter to an ammeter is:

(a) a shunt.
(b) a multiplier.
(c) a thermistor.
(d) a varistor.

5. In a purely capacitive circuit:

(a) the current lags behind the voltage.
(b) the current leads the voltage.
(c) the current and voltage are in step.
(d) the power factor is 20.

6. The scratch filter in a record player is:

(a) a low-pass filter.
(b) an electrolytic capacitor.
(c) an iron-core inductor.
(d) a high-pass filter.

7. A thermistor is used in the yoke to:

(a) eliminate inductive kickback.
(b) compensate for changes in yoke temperature.
(c) increase the Q of the yoke.
(d) increase the resistance of the yoke.

8. When replacing a 100-μF electrolytic capacitor in a power supply filter circuit, you should NOT use:

(a) a 50-μF electrolytic capacitor.
(b) an 80-μF electrolytic capacitor.
(c) a 150-μF electrolytic capacitor.
(d) a 175-μF electrolytic capacitor.

9. The most frequently used types of resistors in radios and television receivers are carbon composition and:

(a) VDR.
(b) thermistor.
(c) wirewound.
(d) ballast.

10. A resistor across a parallel tuned circuit will:

(a) increase its Q.
(b) prevent it from having a resonant frequency.
(c) lower its Q.
(d) decrease the power dissipated by the circuit.

11. The cutoff voltage for an FET is called the:

(a) low gate.
(b) IGFET.
(c) field effect.
(d) pinchoff voltage.

12. To reduce eddy currents, the transformer:

(a) core is laminated.
(b) turns ratio is reduced.
(c) employs windings with lower resistance.
(d) is made with a tunable core.

13. A variable resistor may be connected in either of two ways:

(a) as a VDR and as a thermistor.
(b) as a rheostat and as a potentiometer.
(c) as a peaker and as a nuller.
(d) as a shaper and as a peaker.

14. In most cases, the resistance of a thermistor will:

(a) increase with an increase in temperature.
(b) decrease with an increase in temperature.
(c) be unaffected by changes in temperature.
(d) vary over wide limits with small changes in frequency.

15. A color code of brown black black means a resistance value of:

(a) 1000 ohms.
(b) 100 ohms.
(c) 10 ohms.
(d) 1.0 ohm.

16. Two kinds of electrolytic capacitors are aluminum and:

(a) glass.
(b) tantalum.
(c) paper.
(d) silver mica.

17. To obtain a higher capacitance:

(a) put capacitors in series.
(b) put capacitors in parallel.
(c) use capacitors with a higher voltage rating.
(d) use capacitors with a positive temperature coefficient.

18. The opposition that a capacitor offers to the flow of dc current is called its:

(a) EVR.
(b) ESR.
(c) VSWR.
(d) VDCW.

19. The amount of dc voltage that can be placed across a capacitor continuously without danger of insulation breaking down is:

(a) EVR.
(b) ESR.
(c) VSWR.
(d) VDCW.

20. Which of the following is true about thermistors?

(a) Doubling the voltage across a thermistor doubles the current through it.
(b) Doubling the current through a thermistor doubles its resistance.
(c) The relationship between the voltage across a thermistor and the current through it cannot be determined by Ohm's law.
(d) They are the same thing as VDR's.

21. Which of the following is NOT a linear resistor?

(a) Carbon composition.
(b) Film.
(c) Thermistor.
(d) Wirewound.

22. A mylar capacitor has:

(a) a mylar protective coating.
(b) a mylar dielectric.
(c) a mylar lead.
(d) a very low breakdown voltage.

23. Which of the following is not likely to affect color purity?

(a) The position of the yoke.
(b) Stray magnetic fields.
(c) Adjustment of the brightness control.
(d) Adjustment of the purity magnet.

24. To decrease the brightness of the trace, the cathode of a CRT will be made more:

(a) positive.
(b) negative.

25. The blue lateral correction magnet should be:

(a) at the top or bottom of the gun.
(b) on the right or left side of the gun.

26. Which of the following is the most important consideration when choosing a replacement yoke?

(a) The length of the yoke.
(b) The deflection angle.
(c) The weight of the yoke.
(d) The size of the wire used in the yoke.

27. One feature of a carbon microphone is:

(a) its very low noise characteristic.
(b) its very high impedance when compared to a crystal microphone.
(c) its exceptional high-fidelity characteristics when compared with all other types.
(d) the fact that a dc current must flow through the microphone carbon button in order for it to operate.

28. Which of the following is a component that changes its resistance with changes in light falling on it?

(a) VDR.
(b) LDR.
(c) LTR.
(d) RLS.

29. Coils that are wound on a doughnut-shaped core have a higher inductance per size than other types of inductors. A coil that is wound this way is called:

(a) a doil.
(b) a toroid.
(c) a permut.
(d) an inductoil.

30. An important feature of ceramic capacitors is that they:

(a) all have positive temperature coefficients.
(b) all have zero temperature coefficients.
(c) all have negative temperature coefficients.
(d) can be made with positive, negative, or zero temperature coefficients.

PART IV—TRANSISTORS AND SEMICONDUCTORS

1. Which of the following types of transistors is named because of its appearance (before it is packaged)?

(a) Grown junction.
(b) Alloy.
(c) Unijunction.
(d) Planar.

2. Two examples of breakdown diodes are avalanche diodes and:

(a) zener diodes.
(b) point-contact diodes.
(c) varactor diodes.
(d) ignitrons.

3. In which of the following circuits would you expect to find a varactor diode?

(a) AGC.
(b) AFC.
(c) ABL.
(d) PBY.

4. Which of the following is sometimes used as an oscillator?

(a) A PN junction diode.
(b) A point-contact diode.
(c) A zener diode.
(d) A tunnel diode.

5. A diac is:

(a) a four-layer diode.
(b) a three-layer diode.
(c) an SCR that is reverse biased.
(d) a PN junction.

6. Two types of MOSFET's are depletion and:

(a) washout.
(b) avalanche.
(c) enhancement.
(d) zener.

7. **Compared with a transistor, a MOSFET is a:**

(a) high-impedance device. (b) low-impedance device.

8. **Which of these transistor types has a gate rather than a base?**

(a) NPN.
(b) PNP.
(c) Mesa.
(d) JFET.

9. **A triac is the equivalent of:**

(a) two SCR's.
(b) three diacs.
(c) an ovshinsky diode.
(d) two thyrite resistors.

10. **A pnp transistor is used as an audio voltage amplifier in a certain receiver. The voltage on the collector is negative, and the voltage on the base is positive with respect to the emitter. Which of the following is true?**

(a) This is normal operation for a pnp transistor.
(b) This is normal if the voltage on the emitter is positive.
(c) The voltage on the base is wrong.
(d) The voltage on the collector is wrong.

11. **Moving the plates of a capacitor closer together will:**

(a) increase its capacitance. (b) decrease its capacitance.

12. **A semiconductor device that acts like a voltage-variable capacitor is the:**

(a) MOSFET.
(b) triac.
(c) varactor.
(d) NPO capacitor.

13. **Which of the following can conduct in either of two directions?**

(a) Bipolar transistors.
(b) Npn transistors.
(c) Pnp transistors.
(d) Triacs.

14. **Semiconductor devices that have characteristics similar to those of a thyratron are:**

(a) thyristors.
(b) pnp junction transistors.
(c) MOSFET's.
(d) JFET's.

15. **Which of the following transistor types is especially suited for switching applications?**

(a) Unijunction.
(b) Mesa.
(c) Planar.
(d) Pnp junction type.

16. **Another name for Esaki diode is:**

(a) n-channel JFET.
(b) four-layer diode.
(c) PIN diode.
(d) tunnel diode.

17. **A semiconductor device that has a constant voltage across it for different current values is the:**

(a) point-contact diode.
(b) pn junction diode.
(c) zener diode.
(d) IGFET.

18. **Which of the following accounts for the fact that silicon diodes can handle high forward currents without overheating?**

(a) They have a high peak-inverse voltage.
(b) They are manufactured with built-in heat sinks.
(c) They have a positive temperature coefficient.
(d) They have a low forward resistance.

19. Thermistors are used in transistor circuits to compensate for:

(a) temperature changes.
(b) voltage changes.
(c) current changes.
(d) excessively cold collector junctions.

20. A static charge on a meter probe is liable to destroy a:

(a) varactor.
(b) zener.
(c) MOSFET.
(d) npn junction transistor.

21. Which of the following may be used in place of a varactor?

(a) An IGFET.
(b) A triac.
(c) A pnp transistor collector junction.
(d) The emitter junction of a mesa transistor.

22. Which of these semiconductor devices has a characteristic similar to a neon lamp, except that it *fires* at a much lower voltage?

(a) An n-channel JFET.
(b) A four-layer diode.
(c) An Esaki diode.
(d) A PIN diode.

23. Which of the following types of transistors has no collector?

(a) Npn junction type.
(b) Planar.
(c) Mesa.
(d) Unijunction.

24. Comparing the emitter-base voltage of germanium and silicon transistors:

(a) this voltage is normally lower for a germanium transistor than it is for a silicon transistor.
(b) this voltage is normally higher for a germanium transistor than it is for a silicon transistor.

25. In order to assure a good thermal contact between a power transistor and its heat sink, use:

(a) a 'C' clamp.
(b) silicone grease.
(c) a drop of machine oil.
(d) aguadag.

26. Which of the following is most like a four-layer diode?

(a) An npn transistor direct coupled to a pnp transistor.
(b) A zener diode in series with a triac.
(c) A diac in series with a triac.
(d) An SCR.

27. The characteristic curve for a junction diode is shown in:

(a) Fig. 12-3A.
(b) Fig. 12-3B.
(c) Fig. 12-3C.
(d) Fig. 12-3D.

28. The characteristic curve of a tunnel diode is shown in:

(a) Fig. 12-3A.
(b) Fig. 12-3B.
(c) Fig. 12-3C.
(d) Fig. 12-3D.

29. **The characteristic curve of a transistor (collector voltage vs collector current) is shown in:**

(a) Fig. 12-3A.
(b) Fig. 12-3B.
(c) Fig. 12-3C.
(d) Fig. 12-3D.

30. **The characteristic curve of a zener diode is shown in:**

(a) Fig. 12-3A.
(b) Fig. 12-3B.
(c) Fig. 12-3C.
(d) Fig. 12-3D.

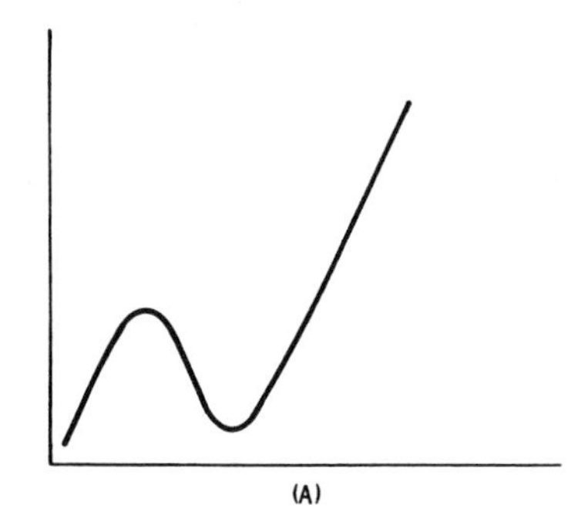
(A)

(B)

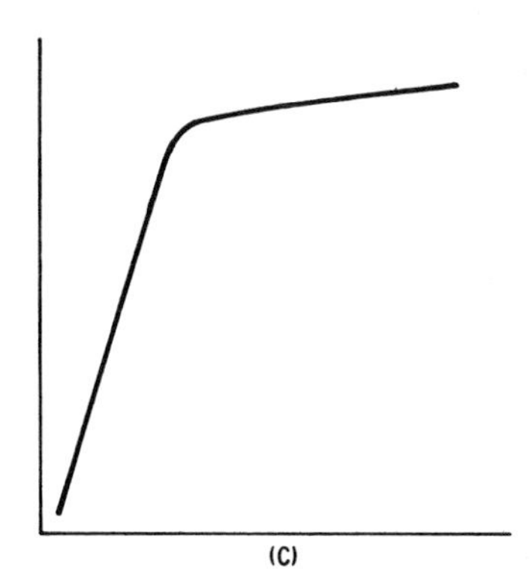
(C)

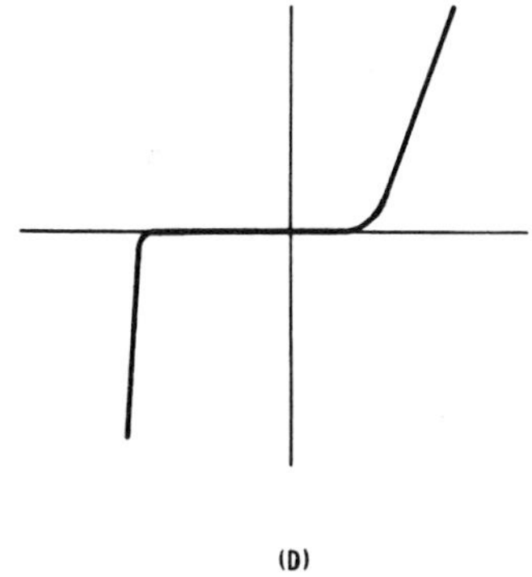
(D)

Fig. 12-3. Characteristic curves.

PART V—BASIC MATHEMATICS AND CIRCUIT ANALYSIS

1. **Two 500-ohm resistors are in series across an ac circuit. The applied frequency is 60 hertz. The impedance of the series circuit is:**

(a) zero ohms.
(b) 1K ohms.
(c) 250 ohms.
(d) 1.57 megohms.

2. **A variable resistor is connected so that it regulates the circuit current. This connection is called a:**

(a) rheostat.
(b) decade.
(c) potentiometer.
(d) divider.

3. The reciprocal of resistance is:

(a) susceptance.
(b) conductance.
(c) admittance.
(d) vars.

4. Fig. 12-4 shows a power triangle. The hypotenuse of this triangle—that is, line AB—represents:

(a) the power dissipated by heat.
(b) the apparent power.
(c) the power that is "borrowed" by an inductor or capacitor.
(d) the power factor.

5. Two coils are wound on identical coil forms with the same number of turns. On one coil the turns are close together, and on the other they are spaced further apart. Which coil will have the greater inductive reactance at 100 Hz?

(a) The coil with the closely spaced turns.
(b) The coil with the widely spaced turns.

6. Normally, the ac voltage reading by a vom is the:

(a) peak-to-peak value.
(b) peak value.
(c) average value.
(d) rms value.

7. When two capacitors are placed in series across an ac circuit:

(a) the larger voltage drop will be across the capacitor with the larger capacitance value.
(b) the larger voltage drop will be across the capacitor with the smaller capacitance value.

8. In a purely capacitive circuit:

(a) the current and voltage are in phase.
(b) the current and voltage are 90° out of phase.
(c) the current and voltage are 180° out of phase.
(d) the voltage across the capacitor will not vary.

9. The time constant of an RC circuit is the time it takes for the capacitor to discharge through the resistor to:

(a) 63 percent of its fully charged voltage.
(b) 37 percent of its fully charged voltage.
(c) 0 volts.
(d) one-half of its fully charged voltage.

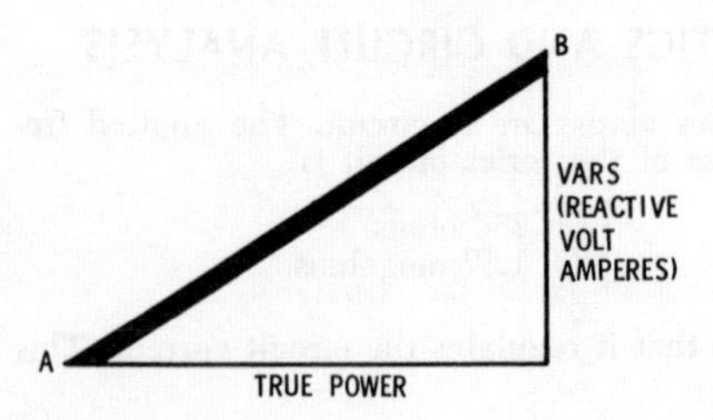

Fig. 12-4. A power triangle.

10. **Which of the following is not true about parallel resonant circuits?**

(a) They can be tuned with a variable resistor in the L or C branch.
(b) It is possible to have a parallel-tuned circuit that does not have a resonant frequency.
(c) Increasing the capacitance increases the resonant frequency.
(d) Impedance is maximum at the resonant frequency.

11. **A certain capacitor is rated at 100 volts. Can this capacitor be used across a 120-volt ac line?**

(a) Yes. (b) No.

12. **How will the Q of a coil be affected when the signal frequency is increased from 1 kHz to 2 kHz?**

(a) It will increase.
(b) It will decrease.
(c) It will stay the same.

13. **A resistor is color coded brown black red gold. Its exact resistance value is 980 ohms. Is this resistance in tolerance?**

(a) Yes. (b) No.

14. **Consider the circuit of Fig. 12-5. Which of the following is true?**

(a) The voltage between A and B is 4 V.
(b) The voltage between A and B is 1.0 V.
(c) The voltage between A and B is 6 V.
(d) The voltage between A and B cannot be determined.

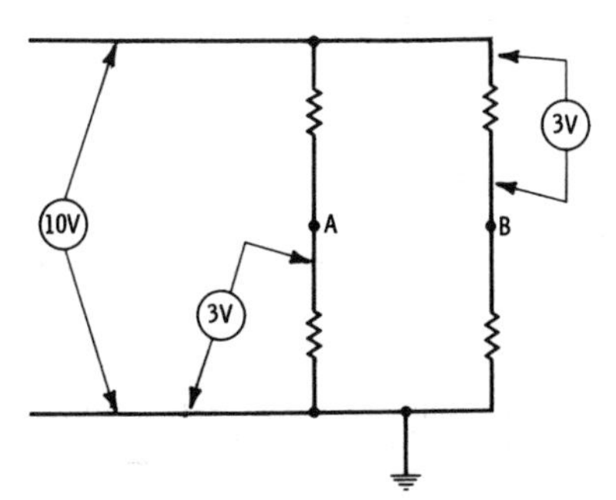

Fig. 12-5. Kirchoff's voltage law can be used for an easy solution of this problem.

15. **Conductance is measured in:**

(a) vars.
(b) mhos.
(c) ohms.
(d) darafs.

16. **When the capacitive reactance in a series RC circuit is 40 ohms, and the resistance value is 30 ohms, then the impedance is:**

(a) 70 ohms.
(b) 10 ohms.
(c) 50 ohms.
(d) 1200 ohms.

17. **The time constant of an LR circuit is given by the equation:**

(a) $T = LR$.
(b) $T = R/L$.
(c) $T = L + R$.
(d) $T = L/R$.

18. The higher the Q of a tuned circuit:

(a) the greater the bandwidth.
(b) the narrower the bandwidth.
(c) the lower the voltage across the capacitor.
(d) the lower the noise.

19. If you charge a capacitor to 100 volts and then move its plates apart:

(a) the voltage across the capacitor will decrease.
(b) the voltage across the capacitor will increase.
(c) the voltage across the capacitor will not change.
(d) the capacitor charge will be completely removed.

20. When I = 0.1 mA and E = 10 V, then according to Ohm's law, R = :

(a) 1K ohms.
(b) 10K ohms.
(c) 100K ohms.
(d) 1 megohm.

21. The average value of a sine-wave voltage that has an rms value of 100 volts is:

(a) 63.6 V.
(b) 50 V.
(c) 31.8 V.
(d) 90 V.

22. Which of these equations is not correct for dc?

(a) $P = E^2/R$.
(b) $P = EI$.
(c) $P = I^2/R$.
(d) $E = IR$.

23. When 0.3 milliampere flows through a 10K resistor, the voltage drop across the resistor is:

(a) 0.3 V.
(b) 3 V.
(c) 30 V.
(d) 0.03 V.

24. Resistors are added to tuned circuits in order to:

(a) raise their Q.
(b) make them tune more sharply.
(c) reduce their noise.
(d) broaden their tuning response.

25. The sum of the powers dissipated by resistors in a series-resistance dc circuit is the total circuit power dissipation. When resistors are in parallel, the total power dissipated by the circuit is:

(a) less than the power dissipated by the smallest resistor.
(b) some value between the power dissipated by the largest and the smallest resistance values.
(c) the smallest value of power in the circuit.
(d) the sum of all the powers dissipated by the resistors.

26. The Q of a coil is determined by the equation:

(a) $Q = L/R$.
(b) $Q = LR$.
(c) $Q = X_L/R$.
(d) $Q = X_LR$.

27. The emitter, base, and collector dc currents of a transistor are I_E, I_B, and I_C respectively. Which of the following is correct?

(a) $I_C = I_B - I_E$.
(b) $I_C = I_E \times I_B$.
(c) $I_E = I_C \times I_B$.
(d) $I_E = I_C + I_B$.

28. Capacitive reactance:

(a) is measured in farads.
(b) increases with an increase in frequency.
(c) is independent of capacitance.
(d) decreases at a given frequency if the plates of a capacitor are moved closer together.

29. A nanosecond is:

(a) 1000 milliseconds.
(b) 10,000 microseconds.
(c) 10^{-9} seconds.
(d) 10^{-12} seconds.

30. Gain-bandwidth product is:

(a) a frequency.
(b) a voltage.
(c) a current.
(d) a power.

PART VI—MONOCHROME TELEVISION

1. A keystone raster:

(a) is necessary for some of the newer shapes in picture tubes.
(b) occurs when the sync pulses are rounded because of the poor high-frequency response of the receiver circuits.
(c) occurs when the yoke is defective.
(d) occurs when the horizontal sweep of the receiver is not linear.

2. The video peaker control is used to:

(a) set the amplitude of the white level in the video amplifier.
(b) set the frequency response of the video amplifier.
(c) set the amplitude of the sync pulses delivered to the video amplifier.
(d) reduce the buzz in the speaker.

3. Two amplifiers are arranged in a circuit so that the same dc current is through both. These amplifiers are said to be:

(a) cascaded.
(b) stacked.
(c) direct coupled.
(d) inverted.

4. The brightness control may be located in a video amplifier stage when:

(a) the manufacturer wants to cut costs.
(b) the picture tube is driven at the cathode rather than at the grid.
(c) keyed agc is used.
(d) the video amplifiers are direct coupled to the picture tube.

5. Which of the following must be a nonlinear stage?

(a) The audio driver.
(b) The i-f amplifier.
(c) The rf amplifier.
(d) The mixer.

6. While earlier model television receivers usually employed a 21-MHz sound i-f, most modern receivers have a sound i-f of:

(a) 10.7 MHz.
(b) 45.5 MHz.
(c) 455 kHz.
(d) 4.5 MHz.

7. In a transistor television receiver, the stage following the video detector is often an emitter follower. The purpose behind using an emitter follower is that it has:

(a) a low input impedance and a high output impedance.
(b) a high input impedance and a low output impedance.

8. A ferrite bead is sometimes used as:

(a) an inductor. (c) a transformer.
(b) a capacitor. (d) a video detector.

9. In a keyed-agc system, the keying pulse comes from:

(a) the video detector. (c) the vertical oscillator.
(b) the video amplifier. (d) the flyback transformer.

10. Which of the following can be used to reduce the problem of Miller effect in a tube circuit?

(a) Use a notch filter.
(b) Use a grounded-grid configuration.
(c) Use an unbypassed cathode resistor.
(d) Use sharp cutoff tubes.

11. A control that sets the level of the video signal near the base of the blanking pedestal, and thus makes it possible to achieve the maximum possible contrast range, is the:

(a) brightness control. (c) agc control.
(b) white-level control. (d) buzz control.

12. Neutralization may be employed in:

(a) i-f amplifier stages. (c) the audio power amplifier.
(b) the tuner local oscillator. (d) the vertical blocking oscillator.

13. A limiter stage is used with:

(a) ratio detectors. (c) pre-emphasis circuits.
(b) discriminators. (d) de-emphasis circuits.

14. An advantage of using pentodes rather than triodes in the receiver i-f stage is that:

(a) they do not generate as much noise.
(b) they do not need to be neutralized.
(c) they can be agc controlled.
(d) they are simpler to service.

15. Two types of agc circuits in transistor television receivers are:

(a) polarized and nonpolarized. (c) electron flow and hole flow.
(b) forward and reverse. (d) conventional and inverted.

16. If a grounded-cathode triode rf amplifier section is not neutralized, the circuit may:

(a) not have sufficient gain. (c) produce static in the speaker.
(b) oscillate. (d) overload the agc circuit.

17. At what point in the receiver is the sound i-f frequency developed?

(a) The mixer-oscillator stage in the tuner.
(b) The noise cancelling circuit.
(c) The video-detector circuit.
(d) The afc circuit.

18. In which section of the television receiver would you expect the amplifier(s) to have the greatest bandwidth?

(a) The video i-f section.
(b) The sound i-f section.
(c) The video amplifier.
(d) The rf amplifier.

19. The range of video frequencies is:

(a) 0-4 MHz.
(b) 1.25-4.5 MHz.
(c) 0-12.5 kHz.
(d) 15-20,000 Hz.

20. In a regulated power supply, you would expect to find:

(a) a tunnel diode.
(b) a zener diode.
(c) a varactor.
(d) a point-contact diode.

21. Normally, you would expect the sync takeoff point to be:

(a) in the video i-f section.
(b) in the tuner.
(c) in the flyback.
(d) in the video detector or video amplifier stages.

22. Traps are used in the i-f stage to:

(a) reduce the amount of interference from radiating local oscillators in the area.
(b) keep the receiver local-oscillator signal out of the i-f section.
(c) shape the i-f response curve.
(d) prevent the sync pulses from getting through the i-f amplifier.

23. The agc voltage to the tuner goes to:

(a) a varactor diode.
(b) the mixer stage.
(c) the local oscillator.
(d) the rf amplifier.

24. In a television receiver, as with any superheterodyne, a low i-f frequency provides a higher receiver gain, but a higher i-f frequency provides

(a) less noise.
(b) better sound quality.
(c) a lower-cost i-f stage.
(d) better image rejection.

25. A differentiator is associated with:

(a) the audio section.
(b) the vertical sync section.
(c) the horizontal sync section.
(d) the video amplifier stage.

26. To reduce the possibility of getting sound in the picture:

(a) the speaker is located as far as possible from the picture tube.
(b) a 4.5-MHz trap is employed in the video amplifier section.
(c) the sound frequencies are separated from the video frequencies by at least 20 MHz.
(d) the sound is located at the maximum point on the response curve and the video is located at the minimum point.

27. A radio receiver in which an i-f amplifier also serves as an audio amplifier is:

(a) a reflex receiver.
(b) a netrodyne receiver.
(c) impossible to design.
(d) a trf receiver.

28. The uhf tuner usually has:

(a) an rf amplifier, an oscillator, and a mixer.
(b) a mixer and local oscillator, but no rf amplifier.
(c) an emitter-follower rf amplifier.
(d) a MOSFET rf amplifier.

29. You would expect to find a gated-beam detector in:

(a) the video section.
(b) the sound section.

30. An agc system that permits the receiver to amplify weak signals with full maximum gain without an agc voltage, but presents an agc control voltage to reduce receiver gain on strong signals, is called:

(a) backward agc.
(b) signal-derived agc.
(c) delayed agc.
(d) agc bias pack.

PART VII—COLOR TELEVISION

1. If varying the color control of a color receiver causes the flesh tones to change in color:

(a) the color circuits of the receiver are operating normally.
(b) the burst signal amplitude is too low.
(c) the chroma circuits may need alignment.
(d) there is an open circuit in the color demodulator section.

2. Which of the following establishes the saturation of the color picture?

(a) The phase difference between I and Q.
(b) The amplitude of the color signal.
(c) The amplitude of the burst signal.
(d) The frequency of the burst signal.

3. The signal to the fine-tuning indicator circuit comes from:

(a) the local oscillator.
(b) the mixer.
(c) a video i-f amplifier.
(d) the flyback transformer.

4. In order to be able to reproduce any color, you must be able to provide proper amounts of:

(a) voltage, current, and resistance.
(b) hue, saturation, and brightness.
(c) amplitude, phase, and frequency.
(d) time, amplitude, and contrast.

5. In order to receive an NTSC color signal, the receiver must have a three-gun color picture tube.

(a) True.
(b) False.

6. With a synchroscope:

(a) you can lock onto the burst directly.
(b) you cannot see the burst.

7. The hue control should be adjusted for:

(a) proper green for the grass.
(b) proper blue for the sky.
(c) proper ratio between black and white.
(d) proper skin tones.

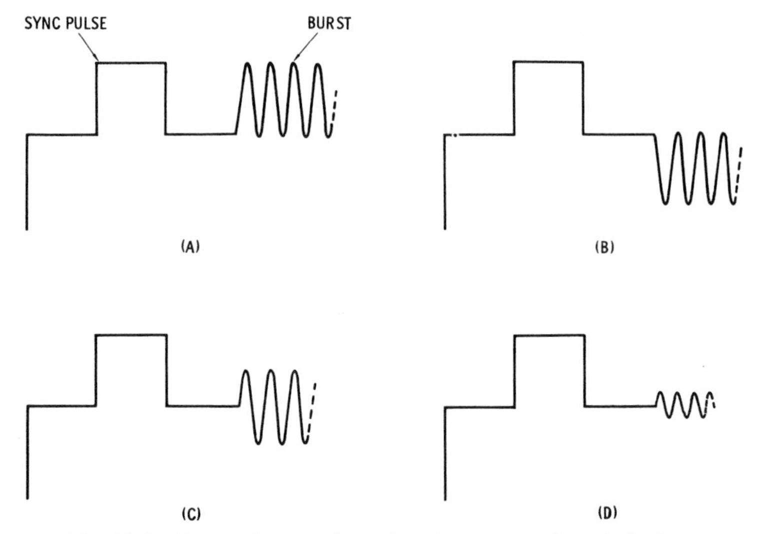

Fig. 12-6. Choose the waveform that shows a portion of the burst signal properly.

8. In Fig. 12-6, which waveform shows the burst as it would be seen at the video detector?

(a) The one shown in A.
(b) The one shown in B.
(c) The one shown in C.
(d) The one shown in D.

9. The circuit that determines whether a signal will pass through the color section is the:

(a) color oscillator.
(b) color killer.
(c) delay line.
(d) color phase.

10. In the circuit of Fig. 12-7, which variable resistor is used for color control?

(a) R1.
(b) R2.
(c) neither.

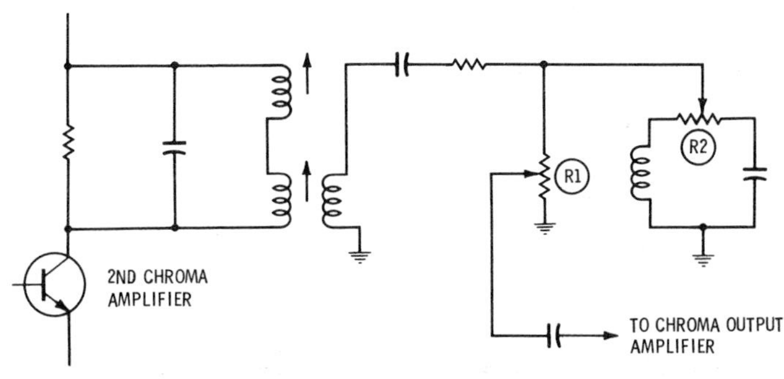

Fig. 12-7. Identify the controls used in this circuit.

11. In the circuit of Fig. 12-7, which variable resistor is the tint control?

(a) R1.
(b) R2.
(c) neither.

12. When signals are in quadrature they are:

(a) in phase.
(b) 90° out of phase.
(c) 180° out of phase.
(d) equal in amplitude.

13. Background controls set:

(a) the automatic fine-tuning average voltage.
(b) the dc bias levels at the color picture tube.
(c) the maximum amplitude of the video signal.
(d) the noise level in the sound section.

14. A correction voltage from the aft circuit will change the:

(a) color-oscillator frequency.
(b) local-oscillator frequency.
(c) chroma-amplifier gain.
(d) amount of delay introduced by the delay line.

15. The brightness control in a color receiver:

(a) can be adjusted with the color control turned off.
(b) cannot be located in the stages that amplify the video.
(c) can eliminate the color sync signals if improperly adjusted.
(d) sets the value of high voltage at the color picture tube.

16. A demodulator that utilizes a reference signal which has the same frequency as the carrier or subcarrier of the signal being demodulated is:

(a) a slope detector.
(b) a synchronous demodulator.
(c) a gated-beam detector.
(d) square-law demodulator.

17. The human eye is most responsive to the color:

(a) red.
(b) blue.
(c) green-yellow.
(d) violet.

18. Which of the following is correct?

(a) $E_Y = 0.3E_R + 0.59E_G + 0.11E_B$.
(b) $E_Y = I + 0.9Q$.
(c) $I = 0.9Q$.
(d) $E_R = E_G = E_B$.

19. The reference white in color television is called:

(a) Illuminant A.
(b) Illuminant B.
(c) Illuminant C.
(d) Illuminant D.

20. The delay line operates on the:

(a) I signal.
(b) Q signal.
(c) burst signal.
(d) luminance signal.

21. The local oscillator frequency of a color receiver is varied by changing the:

(a) acc adjustment.
(b) tint control.
(c) contrast control.
(d) fine-tuning control.

22. The output signal from the burst amplifier will most likely go:

(a) to the color killer and phase detector.
(b) directly to the 3.58-MHz oscillator.
(c) to the color demodulators.
(d) to the blanker.

23. The amplitude of the color signal delivered to the color demodulators is varied by varying the:

(a) hue control.
(b) color control.
(c) contrast control.
(d) picture-tube bias controls.

24. For correct red purity, a microscope should show:

(a) only the green and blue dots illuminated.
(b) only the red dots illuminated.
(c) all three color dots illuminated.
(d) equal brightness across the face of the tube.

25. The technique of making three beams in a color picture tube converge as the beams are swept back and forth is called:

(a) dynamic convergence.
(b) static convergence.
(c) loading.
(d) purity.

26. The automatic color-control adjustment determines the maximum amount of signal that can be delivered to the demodulators. This adjustment sets:

(a) the amplitude of the burst signal.
(b) the gain of the luminance channel.
(c) the gain of the chroma amplifier.
(d) the time duration of the keying pulse.

27. The I and Q signals:

(a) are 90° out of phase.
(b) have the same frequency and are 180° out of phase.
(c) are always in phase.
(d) combine to regulate the frequency of the 3.58-MHz oscillator.

28. In order to accommodate the color signal, the 15,750-Hz horizontal frequency was:

(a) lowered slightly.
(b) raised slightly.
(c) not changed.
(d) doubled.

29. Beam-landing control:

(a) sets the dynamic convergence voltage when the beam is at the center of the screen.
(b) is usually accomplished by adjusting the purity magnets and by positioning the yoke.
(c) is accomplished by properly setting the brightness and contrast controls.
(d) is accomplished by properly setting the linearity adjustment.

30. The signal from the 3.58-MHz oscillator will go to:

(a) the burst amplifier.
(b) a chroma amplifier.
(c) the color demodulators.
(d) the blanker.

PART VIII—SYNCHRONIZING CIRCUITS

1. The damper circuit:

(a) prevents ringing.
(b) does not operate if there is no signal present.
(c) protects the high voltage circuits from fire.
(d) is not necessary when the vertical sweep circuit employs retrace blanking.

2. Which of the following is true?

(a) The left half of the picture is scanned during damper conduction.
(b) The right half of the picture is scanned during damper conduction.

3. To avoid meter damage, do not measure:

(a) the gate voltage on a MOSFET.
(b) the boosted B^+ voltage.
(c) transistor voltages when they are connected into pulse circuits.
(d) the plate voltage on the horizontal output stage.

4. The width of the raster can be varied by changing the height of the sync pulse.

(a) True.
(b) False.

5. The total number of lines transmitted per second in a color picture is slightly less than:

(a) 30.
(b) 60.
(c) 525.
(d) 15,750.

6. A yoke can be tested:

(a) by measuring its resistance with an ohmmeter.
(b) using an oscilloscope with a dual trace.
(c) with a ringing test.
(d) by connecting it across the power line and watching for smoke.

7. The reason that the plate cap on the horizontal output tube should not be disconnected is that:

(a) excessive screen current may flow in the tube.
(b) control-grid current may flow in the tube.
(c) the afc tube may burn out.
(d) the horizontal oscillator may become overloaded.

8. The vertical-linearity control usually affects:

(a) the top half of the picture.
(b) the bottom half of the picture.

9. Afc is used in the:

(a) vertical-oscillator circuit.
(b) horizontal-oscillator circuit.

10. When the vertical-oscillator frequency is too low:

(a) the picture will roll up.
(b) the picture will roll down.

11. To get a sawtooth (or triangular) current wave to flow through a pure inductance, which type of voltage should be placed across the coil?

(a) Sawtooth.
(b) Square wave.
(c) Sine wave.
(d) Parabolic wave.

12. Transistor receivers do not have:

(a) flyback transformers.
(b) dampers.
(c) high-voltage rectifiers.
(d) a problem with Barkhausen oscillations.

13. Which of the following could not be used to synchronize a multivibrator?

(a) A pulse.
(b) A sine wave.
(c) A sawtooth wave.
(d) The output from an electronic filter in the low-voltage supply.

14. To straighten a picture that is tilted:

(a) rotate the tube until the picture is straight.
(b) decrease the high voltage.
(c) rotate the yoke until the picture is straight.
(d) place blocks under the chassis on the side that is too low.

15. If you have just replaced the high-voltage regulator, it would be a good idea to check the:

(a) high voltage.
(b) sync-separator input-signal voltage.
(c) yoke current.
(d) video-detector output-signal voltage.

16. Which of the circuits shown in Fig. 12-8 is a differentiating circuit?

(a) The circuit shown in Fig. 12-8A.
(b) The circuit shown in Fig. 12-8B.

17. Capacitance may be added to the damper-tube circuit in order to:

(a) decrease the brightness range.
(b) correct width or horizontal-linearity problems.
(c) reduce the danger of fire.
(d) protect the picture tube from arcing.

18. Normally, the operating frequency of a sweep oscillator is:

(a) slightly above the free-running frequency.
(b) slightly below the free-running frequency.
(c) at least twice the free-running frequency.
(d) about one-half the free-running frequency.

19. Relaxation oscillator circuits used in the sweep sections are the multivibrator and the:

(a) neon sawtooth generator.
(b) phase-shift oscillator.
(c) Hartley oscillator.
(d) blocking oscillator.

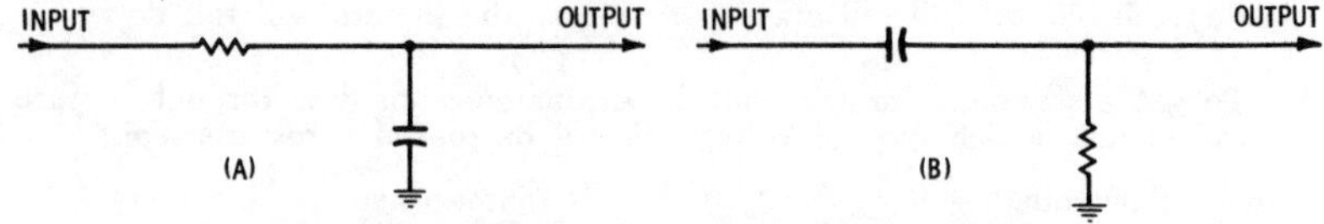

Fig. 12-8. RC networks used in the sync circuits.

20. Which of the following would not receive a signal from the blanker?

(a) The color bandpass amplifier.
(b) The burst amplifier.
(c) The video output stage if it is direct-coupled to the picture tube.
(d) The sweep oscillators.

21. In a certain transistor sync-generator circuit, the emitter and base are at the same voltage. In the absence of an input signal, the transistor is:

(a) saturated.
(b) operating with a normal collector current.
(c) cut off.

22. The picture rolls vertically and horizontally. The trouble is:

(a) in the yoke.
(b) in the picture tube.
(c) in the contrast control.
(d) in the sync separator.

23. Which of the following statements is not correct?

(a) There are two fields per frame.
(b) There are two frames per field.

24. Which of the following procedures is preferred as a step in determining if the high-voltage section is operating?

(a) Pull the horizontal output tube out of the socket and see if there is an arc at the base.
(b) See if the high-voltage lead will arc to the chassis.
(c) Place a temporary short across the damper.
(d) Measure the boost voltage.

25. In a color receiver there is a signal coupled from the horizontal yoke to the vertical yoke. The purpose of this is:

(a) to increase the height of the picture.
(b) to synchronize the vertical sweep to the horizontal sweep.
(c) for top and bottom pincushion correction.
(d) for interlacing the scanning.

26. In a tube-type receiver the boost voltage is low. When the yoke is disconnected, the voltage rises to normal. Which is true?

(a) This is an indication that the damper is shorted.
(b) This is an indication that the yoke may be defective.

27. Changes in yoke temperature are compensated for:

(a) by a thermistor.
(b) by an NPO capacitor.
(c) by a VDR.
(d) by a varicap.

28. A raster is:

(a) the mask around the picture tube.
(b) the rectangular pattern of light on the screen.
(c) the complete high-voltage section.
(d) the yoke assembly.

29. A vertical white line in the center of the picture is most likely caused by:

(a) a defective yoke.
(b) a loss of the horizontal sync pulse.
(c) a loss of horizontal oscillation.
(d) an open filament in the horizontal output stage.

30. In a certain circuit, a tube is being used as a switch to discharge a capacitor. The plate of the tube is connected to one capacitor plate and the cathode is connected to the other capacitor plate. The capacitor will be discharged by:

(a) a negative pulse to the grid.
(b) a positive pulse to the grid.

PART IX—TROUBLESHOOTING TECHNIQUES

1. The color bandpass amplifier should pass:

(a) 3.58 MHz above and below the color subcarrier.
(b) 0.5 MHz above and below the color subcarrier.
(c) 0.5 MHz above and 4.2 MHz below the color subcarrier.
(d) only the color subcarrier.

2. To look at the i-f signal, you should use:

(a) a high-capacitance probe.
(b) a low-capacitance probe.
(c) a direct probe.
(d) a detector probe.

3. When using an oscilloscope for measuring voltage, it is calibrated to read the:

(a) average value.
(b) rms value.
(c) peak value.
(d) peak-to-peak value.

4. Which of the following sections in a color television is never aligned?

(a) The tuner.
(b) The i-f section.
(c) The color bandpass amplifiers.
(d) The luminance section.

5. To measure the total current drawn by a small transistor portable radio:

(a) put the milliammeter across the ON-OFF switch terminals and turn the radio OFF.
(b) put the milliammeter across the ON-OFF switch terminals and turn the radio ON.
(c) remove batteries and replace with a voltmeter.
(d) place the milliameter across the power supply.

6. Which of the following will be the result of a loss of color sync?

(a) Vertical roll of the picture.
(b) Picture out of horizontal hold.
(c) Loss of color.
(d) Improper color.

7. For checking a receiver by signal tracing rather than by signal injection:

(a) you start at the output and work toward the antenna.
(b) you start at the antenna and work toward the output.
(c) you use a sweep generator rather than an rf generator.
(d) the receiver is OFF instead of ON.

8. The standard i-f frequency for a table-model a-m broadcast receiver is:

(a) 10.7 MHz.
(b) 455 kHz.
(c) 3.58 MHz.
(d) 45.75 MHz.

9. In a pnp transistor amplifier, the base voltage should be:

(a) positive with respect to the emitter voltage.
(b) positive with respect to the collector voltage.

10. A radio receiver is tuned to a weak station. Which of the following is correct?

(a) The base voltage of the mixer stage should change when a strong station is tuned in.
(b) The base voltage of the mixer stage should not change when a strong station is tuned in.

11. Do not test transistors with an ohmmeter set on the:

(a) R × 1 scale.
(b) R × 1K scale.
(c) R × 10K scale.
(d) R × 100K scale.

12. The adjacent-channel sound trap is usually adjusted to:

(a) 4.5 MHz.
(b) 45.75 MHz.
(c) 47.25 MHz.
(d) 10 kHz.

13. The last step in a color-TV setup procedure is:

(a) high-voltage adjustment.
(b) purity adjustment.
(c) gray-scale adjustment.
(d) static convergence.

14. The frequency in most sweep generators is altered at a rate of:

(a) 30 Hz.
(b) 60 Hz.
(c) 120 Hz.
(d) 400 Hz.

15. If you suspect that the local oscillator of a receiver is not producing a signal, you can check it using:

(a) a sweep generator.
(b) the radiated signal from a local oscillator in another receiver.
(c) a screwdriver to draw an arc from the oscillator coil.
(d) a dc voltage across the oscillator coil.

16. The hue control should be:

(a) fully clockwise when the proper flesh tones are displayed.
(b) in the center of its adjustment when the proper flesh tones are displayed.
(c) fully counterclockwise when the proper flesh tones are displayed.
(d) adjusted before the receiver fine-tuning control is adjusted.

17. Leakage current in the transistor collector-base junction when it is reverse biased is called:

(a) I_{CBO}.
(b) I_C.
(c) I_B.
(d) I_{EOC}.

18. Which is normally larger?

(a) The bias on an npn silicon transistor.
(b) The bias on an npn germanium transistor.

19. The standard i-f frequency for a table-model fm broadcast receiver is:

(a) 10.7 MHz.
(b) 455 kHz.
(c) 3.58 MHz.
(d) 45.75 MHz.

20. Positive-going pulses at the base of a Class-A common-emitter npn amplifier should produce:

(a) negative-going pulses at the collector.
(b) positive-going pulses at the collector.

21. The best way to check a coupling capacitor for leakage is to:

(a) measure the dc voltage across it.
(b) bridge it with another capacitor.
(c) disconnect it and substitute another capacitor known to be good.
(d) place a temporary short across it to see if the signal will go to the next stage.

22. When you use a keyed rainbow generator, the display should show:

(a) blue on the left side of the screen and red on the right side.
(b) green between the red and blue display.
(c) yellow-orange on the left side of the screen and green on the right side.
(d) vertical dark stripes in the complete range of the gray scale.

23. In the normal operation of a transistor, which junction is forward biased?

(a) The emitter-base junction.
(b) The collector-base junction.

24. The forward emitter-base resistance of a silicon transistor measures 200 ohms, and the reverse resistance is 400 ohms. This indicates that:

(a) the transistor is good.
(b) the resistance ratio is too low for a good junction.
(c) the ohmmeter is defective.
(d) the transistor is incorrectly marked. This is obviously the collector-base junction.

25. In the color television setup procedure, which of the following steps should be done first?

(a) Purity adjustment.
(b) Picture size adjustment.
(c) Focus adjustment.
(d) High-voltage adjustment.

26. During reception of a monochrome signal, the color killer cuts off the:

(a) 3.58-MHz oscillator.
(b) luminance amplifier.
(c) matrix section.
(d) bandpass amplifier.

27. In comparing a synchroscope with an oscilloscope:

(a) only the oscilloscope can have a dc input.
(b) only the oscilloscope can be used for displaying Lissajous patterns.
(c) the synchroscope will not lock onto a sine wave.
(d) the synchroscope does not have a trace in the absence of an input signal to be displayed.

28. Improper purity adjustment will affect:

(a) the monochrome picture only.
(b) the color picture only.
(c) both the monochrome and the color picture.
(d) neither the monochrome picture nor the color picture.

29. When measuring the front-to-back resistance of a certain diode, it is found to be 10 to 1. The diode is:

(a) good.
(b) not good.

30. If you are going to use an ohmmeter for testing transistors, how many measurements should you make for the most thorough check?

(a) Two.
(b) Four.
(c) Six.
(d) Eight.

PART X—WAVEFORM ANALYSIS

1. Actually, the daisy pattern on a vectorscope is a form of:

(a) linearity measurement.
(b) gain measurement.
(c) Lissajous pattern.
(d) current measurement.

2. As measured with an oscilloscope, you would expect the signal voltage at the output of the video amplifier to be:

(a) 100 V.
(b) 1.5 V.
(c) 15 millivolts.
(d) 1000 microvolts.

3. In order to display two complete lines of video on the oscilloscope, the sweep frequency of the scope should be set at:

(a) 60 Hz.
(b) 525 Hz.
(c) 7785 Hz.
(d) 15,750 Hz.

4. The widest frequency response in a receiver should be in the:

(a) sound section.
(b) color bandpass amplifier.
(c) video i-f amplifier.
(d) tuner.

5. When two sine-wave signals that are 90° out of phase, but identical in frequency, are compared on an oscilloscope in a Lissajous pattern, the result is:

(a) a straight line.
(b) a sine wave.
(c) an ellipse.
(d) a circle.

6. When setting up the daisy pattern on a vectorscope, the petal that is at the 90° point is:

(a) R – Y.
(b) B – Y.
(c) G – Y.
(d) flesh tone.

7. Absorption traps can be used as markers.

(a) True. (b) False.

8. With the post-marker system of alignment:

(a) the marker cannot be observed on the scope trace when there is no input signal from the receiver.
(b) the markers can be observed on the scope trace even though there is no input signal from the receiver.

9. The waveforms shown in Fig. 12-9 are obtained by feeding a square wave into a video amplifier and observing the output on an oscilloscope.

(a) The waveform in Fig. 12-9A indicates poor high-frequency response.
(b) The waveform in Fig. 12-9B indicates poor high-frequency response.

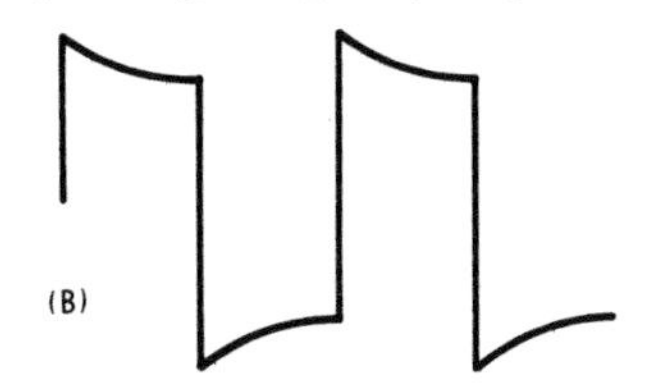

Fig. 12-9. Waveforms obtained when making a square-wave test.

10. In order to use a square-wave generator to check the horizontal linearity of a receiver, its signal is injected into a video amplifier stage, and its frequency is:

(a) less than 60 Hz.
(b) between 60 and 15,750 Hz.
(c) greater than 15,750 Hz.

11. Convergence should be performed in:

(a) a very bright room.
(b) the shop.
(c) a darkened room.
(d) outdoors.

12. The signal delivered to the integrator consists of:

(a) the vertical blanking pedestal and the vertical sync pulses.
(b) the vertical sync pulses, vertical blanking pedestal, and video signal.
(c) The I and Q signals.
(d) none of these.

13. On the i-f response curve of Fig. 12-10, the sound signal should be at point:

(a) A.
(b) B.
(c) C.
(d) D.

14. A figure eight Lissajous pattern indicates that:

(a) one of the frequencies is twice the other.
(b) the scope is not properly grounded.
(c) the vertical input terminal is open.
(d) The internal sweep of the scope is still working.

15. When observing a television i-f response curve on an oscilloscope, the scope sweep is provided by:

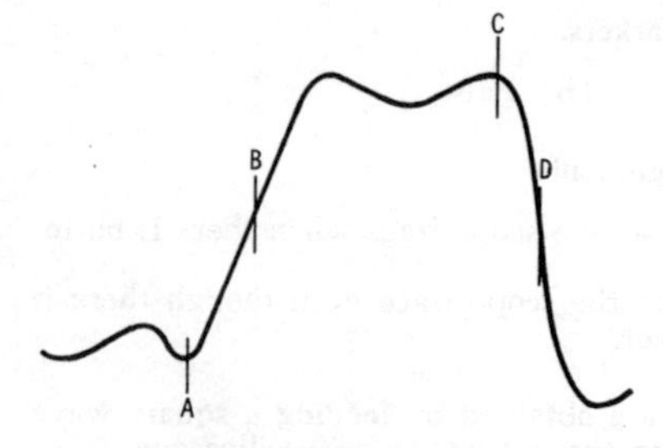

Fig. 12-10. An i-f response curve.

(a) the sweep generator.
(b) the receiver.
(c) the ac power line.
(d) the sawtooth generator in the scope.

16. For the ringing test of a deflection coil, the impulse signal can be obtained from:

(a) the oscilloscope.
(b) a sweep generator.
(c) an absorption trap.
(d) the 3.58 MHz oscillator.

17. In order to look at two waveforms simultaneously, you can use a dual-trace oscilloscope or a conventional oscilloscope and:

(a) a voltage divider probe.
(b) two probes connected to the vertical input.
(c) an electronic switch.
(d) a mixer.

18. The ripple frequency of a bridge rectifier circuit connected across a 400-Hz power line is:

(a) 400 Hz.
(b) 120 Hz.
(c) 800 Hz.
(d) 60 Hz.

19. The method of checking the response curve of a bandpass amplifier with a marker generator and a sweep generator that sweeps 6 MHz is called:

(a) vom.
(b) vsr.
(c) vswr.
(d) vsm.

20. Which of the following is in the *blacker-than-black* region?

(a) Blanking pedestals.
(b) Sync pulses.
(c) Zero signal reference.
(d) VITS.

21. A defective power-supply filter capacitor can be located with:

(a) a sweep generator.
(b) a marker generator.
(c) an oscilloscope.
(d) a marker adder.

22. The color burst is located:

(a) on the front porch of the vertical blanking pedestal.
(b) on the back porch of the vertical blanking pedestal.
(c) on the front porch of the horizontal blanking pedestal.
(d) on the back porch of the horizontal blanking pedestal.

23. The amplitude of the color burst is:

(a) about the same as the amplitude of the sync pulses.
(b) about the same as the amplitude of the I signal.
(c) about the same as the amplitude of the Q signal.
(d) about the same amplitude as the blanking pedestal.

24. Maximum signal strength occurs during the:

(a) color burst.
(b) sync tips.
(c) transmission of the signal corresponding to white areas of the picture.
(d) transmission of the signal corresponding to black areas of the picture.

25. For the post-marker method of sweep alignment:

(a) a marker generator is not needed.
(b) a marker adder is needed.
(c) a scope is not needed.
(d) the receiver must be OFF.

26. When aligning a receiver by using the post-marker system, the markers are introduced between:

(a) the first and second i-f stages.
(b) the last video amplifier and detector.
(c) the sweep generator and the receiver.
(d) the video detector and the oscilloscope.

27. An oscilloscope is used to measure a sine-wave voltage, and it is found to be 100 volts peak-to-peak. A voltmeter, when used to measure the same voltage, would show the value to be:

(a) 70.7 V.
(b) 35.35 V.
(c) 50 V.
(d) 100 V.

28. A true square wave can be obtained by combining a sine wave at a fundamental frequency with:

(a) sine waves at odd harmonic frequencies and decreasing amplitudes.
(b) sine waves at even harmonic frequencies and decreasing amplitudes.
(c) sawtooth waves at even harmonic frequencies and increasing amplitudes.
(d) a sine wave at four times the fundamental frequency.

29. In order to get an accurate response curve during sweep alignment, you should:

(a) set the brightness control fully clockwise.
(b) set the color control fully clockwise.
(c) set the agc amplifier gain to its maximum value.
(d) use a dc bias pack.

30. If you apply a pure sine wave to a differentiating circuit, the output will be:

(a) a square wave.
(b) a sawtooth.
(c) a triangular wave.
(d) a sine wave.

Answers to Practice Tests

CHAPTER 2

1. (b) 2. (a) 3. (d) 4. (b) 5. (b)
6. (b) 7. (a) 8. (b) 9. (d) 10. (c)
11. (c) 12. (b) 13. (c) 14. (d) 15. (b)
16. (c) 17. (b) 18. (b) 19. (b) 20. (a)
21. (a) 22. (b) 23. (c) 24. (b) 25. (a)
26. (b) 27. (c) 28. (c) 29. (c) 30. (d)
31. (b) 32. (b) 33. (c) 34. (c) 35. (d)
36. (a) 37. (d) 38. (b) 39. (c) 40. (b)
41. (c) 42. (c) 43. (d) 44. (c) 45. (d)
46. (a) 47. (b) 48. (c) 49. (c) 50. (b)

CHAPTER 3

1. (c) 2. (b) 3. (a) 4. (b) 5. (c)
6. (c) 7. (a) 8. (b) 9. (b) 10. (c)
11. (d) 12. (c) 13. (b) 14. (a) 15. (c)
16. (d) 17. (d) 18. (a) 19. (b) 20. (c)
21. (a) 22. (b) 23. (a) 24. (b) 25. (a)
26. (d) 27. (c) 28. (a) 29. (c) 30. (b)
31. (a) 32. (b) 33. (c) 34. (c) 35. (c)
36. (c) 37. (a) 38. (b) 39. (d) 40. (d)
41. (e) 42. (a) 43. (a) 44. (c) 45. (b)
46. (d) 47. (b) 48. (c) 49. (a) 50. (a)

CHAPTER 4

1. (b) 2. (b) 3. (c) 4. (d) 5. (a)
6. (c) 7. (a) 8. (d) 9. (b) 10. (d)
11. (a) 12. (d) 13. (a) 14. (c) 15. (d)
16. (d) 17. (b) 18. (b) 19. (b) 20. (c)
21. (c) 22. (a) 23. (a) 24. (b) 25. (b)
26. (a) 27. (b) 28. (b) 29. (b) 30. (b)
31. (c) 32. (d) 33. (c) 34. (c) 35. (a)
36. (d) 37. (a) 38. (b) 39. (a) 40. (b)
41. (d) 42. (b) 43. (c) 44. (a) 45. (b)
46. (a) 47. (d) 48. (c) 49. (d) 50. (b)

CHAPTER 5

1. (b)	2. (c)	3. (a)	4. (d)	5. (a)
6. (b)	7. (c)	8. (a)	9. (c)	10. (d)
11. (c)	12. (a)	13. (d)	14. (a)	15. (a)
16. (b)	17. (b)	18. (d)	19. (b)	20. (c)
21. (d)	22. (d)	23. (d)	24. (b)	25. (a)
26. (d)	27. (d)	28. (b)	29. (c)	30. (b)
31. (d)	32. (c)	33. (b)	34. (c)	35. (b)
36. (d)	37. (a)	38. (b)	39. (c)	40. (d)
41. (d)	42. (b)	43. (c)	44. (c)	45. (b)
46. (a)	47. (a)	48. (c)	49. (b)	50. (b)

CHAPTER 6

1. (b)	2. (b)	3. (a)	4. (c)	5. (b)
6. (d)	7. (b)	8. (c)	9. (a)	10. (a)
11. (c)	12. (d)	13. (a)	14. (d)	15. (b)
16. (c)	17. (b)	18. (d)	19. (b)	20. (d)
21. (c)	22. (b)	23. (c)	24. (a)	25. (a)
26. (d)	27. (b)	28. (b)	29. (c)	30. (c)
31. (b)	32. (b)	33. (a)	34. (a)	35. (d)
36. (d)	37. (c)	38. (b)	39. (b)	40. (b)
41. (b)	42. (d)	43. (b)	44. (c)	45. (a)
46. (c)	47. (d)	48. (d)	49. (c)	50. (c)

CHAPTER 7

1. (b)	2. (a)	3. (c)	4. (b)	5. (c)
6. (a)	7. (d)	8. (b)	9. (a)	10. (b)
11. (a)	12. (d)	13. (a)	14. (a)	15. (d)
16. (b)	17. (b)	18. (b)	19. (a)	20. (c)
21. (b)	22. (b)	23. (c)	24. (a)	25. (a)
26. (b)	27. (d)	28. (c)	29. (c)	30. (b)
31. (d)	32. (c)	33. (d)	34. (b)	35. (a)
36. (b)	37. (a)	38. (c)	39. (c)	40. (c)
41. (a)	42. (c)	43. (b)	44. (a)	45. (d)
46. (a)	47. (b)	48. (a)	49. (d)	50. (c)

CHAPTER 8

1. (d)	2. (a)	3. (c)	4. (b)	5. (a)
6. (d)	7. (b)	8. (c)	9. (d)	10. (b)
11. (d)	12. (a)	13. (d)	14. (a)	15. (b)
16. (b)	17. (b)	18. (c)	19. (a)	20. (c)
21. (b)	22. (b)	23. (c)	24. (b)	25. (d)
26. (b)	27. (d)	28. (c)	29. (c)	30. (b)
31. (d)	32. (c)	33. (a)	34. (b)	35. (a)
36. (b)	37. (a)	38. (b)	39. (c)	40. (d)
41. (a)	42. (a)	43. (b)	44. (b)	45. (c)
46. (d)	47. (b)	48. (c)	49. (d)	50. (b)

CHAPTER 9

1. (b) 2. (b) 3. (c) 4. (a) 5. (d)
6. (c) 7. (a) 8. (b) 9. (d) 10. (c)
11. (b) 12. (b) 13. (c) 14. (a) 15. (c)
16. (c) 17. (c) 18. (a) 19. (a) 20. (a)
21. (c) 22. (c) 23. (b) 24. (b) 25. (d)
26. (d) 27. (b) 28. (b) 29. (c) 30. (c)
31. (b) 32. (a) 33. (c) 34. (a) 35. (a)
36. (a) 37. (a) 38. (b) 39. (c) 40. (b)
41. (b) 42. (a) 43. (a) 44. (b) 45. (c)
46. (d) 47. (c) 48. (d) 49. (a) 50. (a)

CHAPTER 10

1. (a) 2. (b) 3. (b) 4. (a) 5. (a)
6. (b) 7. (a) 8. (a) 9. (b) 10. (d)
11. (c) 12. (b) 13. (b) 14. (a) 15. (a)
16. (b) 17. (a) 18. (a) 19. (b) 20. (d)
21. (a) 22. (c) 23. (d) 24. (d) 25. (a)
26. (b) 27. (c) 28. (b) 29. (b) 30. (a)
31. (b) 32. (c) 33. (b) 34. (d) 35. (a)
36. (c) 37. (c) 38. (c) 39. (d) 40. (a)
41. (b) 42. (a) 43. (a) 44. (a) 45. (a)
46. (a) 47. (d) 48. (b) 49. (b) 50. (c)

CHAPTER 11

1. (c) 2. (a) 3. (b) 4. (c) 5. (a)
6. (b) 7. (a) 8. (d) 9. (a) 10. (c)
11. (c) 12. (d) 13. (b) 14. (c) 15. (b)
16. (a) 17. (a) 18. (d) 19. (b) 20. (a)
21. (d) 22. (c) 23. (c) 24. (c) 25. (c)
26. (a) 27. (d) 28. (b) 29. (d) 30. (b)
31. (c) 32. (d) 33. (d) 34. (b) 35. (b)
36. (c) 37. (b) 38. F 39. M 40. I
41. C 42. D 43. H 44. L 45. E
46. J 47. A 48. K 49. B 50. G

CHAPTER 12

Part I—The Television Signal

1. (d) 2. (d) 3. (d) 4. (a) 5. (b)
6. (c) 7. (c) 8. (a) 9. (b) 10. (c)
11. (b) 12. (b) 13. 33,333 14. (c) 15. (a)
16. (b) 17. 25% 18. 75% 19. 12.5% 20. (b)
21. (b) 22. red green blue 23. (b) 24. (b) 25. (a)
26. (b) 27. (d) 28. (c) 29. (d) 30. (c)

Part II—Antennas and Transmission Lines

1. (a)	2. (a)	3. (a)	4. (a)	5. (b)
6. (b)	7. (b)	8. (a)	9. (a)	10. (b)
11. (b)	12. (b)	13. (a)	14. (b)	15. (b)
16. balun	17. (c)	18. head end, distribution	19. 1000	20. electric
21. (d)	22. wave-length	23. (a)	24. (a)	25. (a)
26. (b)	27. (d)	28. (c)	29. (c)	30. (a)

Part III—Electronic Components

1. (d)	2. (d)	3. (b)	4. (a)	5. (b)
6. (a)	7. (b)	8. (a)	9. (c)	10. (c)
11. (d)	12. (a)	13. (b)	14. (b)	15. (c)
16. (b)	17. (b)	18. (b)	19. (d)	20. (c)
21. (c)	22. (b)	23. (c)	24. (a)	25. (a)
26. (b)	27. (d)	28. (b)	29. (b)	30. (d)

Part IV—Transistors and Semiconductors

1. (d)	2. (a)	3. (b)	4. (d)	5. (b)
6. (c)	7. (a)	8. (d)	9. (a)	10. (c)
11. (a)	12. (c)	13. (d)	14. (a)	15. (a)
16. (d)	17. (c)	18. (d)	19. (a)	20. (c)
21. (c)	22. (b)	23. (d)	24. (a)	25. (b)
26. (a)	27. (b)	28. (a)	29. (c)	30. (d)

Part V—Basic Mathematics and Circuit Analysis

1. (b)	2. (a)	3. (b)	4. (b)	5. (a)
6. (d)	7. (b)	8. (b)	9. (b)	10. (c)
11. (b)	12. (a)	13. (a)	14. (a)	15. (b)
16. (c)	17. (d)	18. (b)	19. (b)	20. (c)
21. (d)	22. (c)	23. (b)	24. (d)	25. (d)
26. (c)	27. (d)	28. (d)	29. (c)	30. (a)

Part VI—Monochrome Television

1. (c)	2. (b)	3. (b)	4. (d)	5. (d)
6. (d)	7. (b)	8. (a)	9. (d)	10. (c)
11. (b)	12. (a)	13. (b)	14. (b)	15. (b)
16. (b)	17. (c)	18. (d)	19. (a)	20. (b)
21. (d)	22. (c)	23. (d)	24. (d)	25. (c)
26. (b)	27. (a)	28. (b)	29. (b)	30. (c)

Part VII—Color Television

1. (c)	2. (b)	3. (c)	4. (b)	5. (b)
6. (a)	7. (d)	8. (c)	9. (b)	10. (a)
11. (b)	12. (b)	13. (b)	14. (b)	15. (a)
16. (b)	17. (c)	18. (a)	19. (c)	20. (d)
21. (d)	22. (a)	23. (b)	24. (b)	25. (a)
26. (c)	27. (a)	28. (a)	29. (b)	30. (c)

Part VIII—Synchronizing Circuits

1. (a)	2. (a)	3. (d)	4. (b)	5. (d)
6. (c)	7. (a)	8. (a)	9. (b)	10. (a)
11. (b)	12. (d)	13. (d)	14. (c)	15. (a)
16. (b)	17. (b)	18. (a)	19. (d)	20. (d)
21. (c)	22. (d)	23. (b)	24. (d)	25. (c)
26. (b)	27. (a)	28. (b)	29. (a)	30. (b)

Part IX—Troubleshooting Techniques

1. (b)	2. (d)	3. (d)	4. (d)	5. (a)
6. (d)	7. (b)	8. (b)	9. (b)	10. (a)
11. (a)	12. (c)	13. (c)	14. (b)	15. (b)
16. (b)	17. (a)	18. (a)	19. (a)	20. (a)
21. (c)	22. (c)	23. (a)	24. (b)	25. (d)
26. (d)	27. (d)	28. (c)	29. (b)	30. (c)

Part X—Waveform Analysis

1. (c)	2. (b)	3. (c)	4. (d)	5. (d)
6. (a)	7. (a)	8. (b)	9. (a)	10. (c)
11. (c)	12. (d)	13. (a)	14. (a)	15. (a)
16. (a)	17. (c)	18. (c)	19. (d)	20. (b)
21. (c)	22. (d)	23. (a)	24. (b)	25. (b)
26. (d)	27. (b)	28. (a)	29. (d)	30. (d)

Index